FM 3-05.231

Special Forces Personnel Recovery

JUNE 2003

DISTRIBUTION RESTRICTION:
Distribution authorized to U.S. Government agencies and their contractors only to protect technical or operational information from automatic dissemination under the International Exchange Program or by other means. This determination was made on 5 December 2003. Other requests for this document must be referred to Commander, United States Army John F. Kennedy Special Warfare Center and School, ATTN: AOJK-DT-SF, Fort Bragg, North Carolina 28310-5000.

DESTRUCTION NOTICE:
Destroy by any method that must prevent disclosure of contents or reconstruction of the document.

Headquarters, Department of the Army

Field Manual
No. 3-05.231

FM 3-05.231
Headquarters
Department of the Army
Washington, DC, 13 June 2003

Special Forces Personnel Recovery

Contents

		Page
	PREFACE	iv
Chapter 1	INTRODUCTION TO PERSONNEL RECOVERY	1-1
	Evolution of SF Support to PR	1-2
	Department of Defense Policy	1-5
	USSOCOM Policy	1-5
	Philosophy Behind PR Operations	1-6
	Five Tasks of PR	1-6
	Recovery Considerations	1-8
	PR Spectrum	1-11
	Options and Methods of PR	1-12
Chapter 2	COMMAND, CONTROL, COORDINATION, AND MISSION MANAGEMENT	2-1
	Command Relationships	2-1
	Premission Planning	2-12
	Mission Employment	2-20
	Postmission	2-23
Chapter 3	PERSONNEL RECOVERY FOR SPECIAL FORCES	3-1
	Section I - Unassisted Evasion: Special Forces as the Evader	3-1

DISTRIBUTION RESTRICTION: Distribution authorized to U.S. Government agencies and their contractors only to protect technical or operational information from automatic dissemination under the International Exchange Program or by other means. This determination was made on 5 December 2003. Other requests for this document must be referred to Commander, United States Army John F. Kennedy Special Warfare Center and School, ATTN: AOJK-DT-SF, Fort Bragg, North Carolina 28310-5000.

DESTRUCTION NOTICE: Destroy by any method that must prevent disclosure of contents or reconstruction of the document.

FM 3-05.231

	Page
Unassisted Evasion	3-1
Training	3-3
Planning for Evasion	3-5
Extended Evasion	3-7
Unplanned Assistance During Evasion	3-8
Support to Evaders	3-8
Evasion Aids	3-8
Reporting Own Situation	3-12
Signaling	3-13
Contact Procedures	3-14
Individual Responsibilities	3-15
Section II - Opportune Support to PR	3-16
Diversion From Primary Mission in Support of PR	3-16
Multiple and Follow-On Missions	3-17
Section III - Unilateral Recovery	3-18
Unilateral Recovery Responsibility	3-18
Unilateral CSAR	3-19
Section IV - Joint Recovery Operations	3-19
SF as the Security and Contact Force for a CSARTF	3-20
Exfiltration of Evader Recovered by a UART	3-21
Exfiltration of an Evader Recovered by an RM	3-22
Operation ALLIED FORCE	3-23
Section V - Support to Multinational PR Operations	3-24
SF Support Role	3-24
UAR in the Coalition Environment	3-24
Section VI - Unconventional Assisted Recovery	3-25
Spectrum of UAR Operations	3-26
Specified Tasks	3-26
RT	3-34
UART	3-34
RM	3-36
UARM	3-37
Operation MARKET GARDEN	3-39
UAR Planning	3-42
UAR Assets	3-43

		Page
Chapter 4	**SPECIAL FORCES LIBERATION**	4-1
	DA Operations	4-1
	Liberation Operations	4-1
	SF DA Mission in Support of PR	4-3
Chapter 5	**CIVIL AFFAIRS AND PSYCHOLOGICAL OPERATIONS SUPPORT TO PERSONNEL RECOVERY**	5-1
	CA Support	5-1
	PSYOP Support	5-2
Appendix A	**EVASION PLAN OF ACTION FORMAT**	A-1
Appendix B	**FORWARD OPERATIONAL BASE EVASION PLAN OF ACTION GUIDANCE FORMAT**	B-1
Appendix C	**EVASION PLAN OF ACTION PLANNING CONSIDERATIONS**	C-1
Appendix D	**UNCONVENTIONAL ASSISTED RECOVERY COORDINATION CENTER OPERATIONS**	D-1
Appendix E	**PERSONNEL RECOVERY PLANNER CHECKLISTS AND VOICE MESSAGE TEMPLATES**	E-1
	GLOSSARY	Glossary-1
	BIBLIOGRAPHY	Bibliography-1
	INDEX	Index-1

Preface

This manual provides a doctrinal framework for Special Forces (SF) personnel recovery (PR) operations. It outlines the contribution of SF to the theater PR effort. It also provides an explanation of the SF PR mission tasks, capabilities, limitations, general guidance, and employment techniques. SF PR missions seek to achieve specific, well-defined, and often sensitive results of strategic or operational significance. SF PR missions are conducted in support of their own operations, when directed by the joint task force commander (JTFC) to support a combat search and rescue (CSAR) operation, when the threat to the recovery force is high enough to warrant the conduct of a special operation, and when SF is the only force available. Detailed planning, rehearsals, and in-depth intelligence analysis characterize SF recovery missions.

This manual addresses SF support to PR, which encompasses, but is not limited to, unassisted evasion, opportune recovery, CSAR, joint combat search and rescue (JCSAR), nonconventional assisted recovery (NAR) and unconventional assisted recovery (UAR), and evasion and recovery (E&R). However, this manual mainly focuses on UAR, which is NAR conducted by special operations forces (SOF). It also addresses the duties, responsibilities, establishment, and use of rescue coordination centers (RCCs), the unconventional assisted recovery coordination center (UARCC), and joint search and rescue centers (JSRCs). Chapter 2 contains information on theater rescue centers.

This manual describes UAR and the role of SF as it relates to PR. In addition, it forms the basis for understanding the unique contribution of UAR to PR with the objective of providing common SF doctrine. SF routinely employs unconventional tactics and techniques while conducting covert or clandestine operations unilaterally and with indigenous assistance. The conduct of UAR operations by SF differs from conventional recovery operations in the degree of political risk, operational techniques, independence from friendly support, and dependence on detailed operational intelligence and indigenous assets.

This manual applies to all United States (U.S.) Army SF groups, to include National Guard components.

NOTE: This publication supersedes USAJFKSWCS Publication 525-5-14, 27 January 1999.

The proponent of this manual is the United States Army John F. Kennedy Special Warfare Center and School (USAJFKSWCS). Submit comments and recommended changes to Commander, USAJFKSWCS, ATTN: AOJK-DT-SF, Fort Bragg, NC 28310-5000.

Unless this publication states otherwise, masculine nouns and pronouns do not refer exclusively to men.

Chapter 1

Introduction to Personnel Recovery

PR is the aggregation of military, civil, and diplomatic efforts to recover captured, detained, evading, isolated, or missing personnel from uncertain or hostile environments and denied areas. PR may occur through military action, actions by nongovernmental organizations (NGOs), other U.S. Government-approved actions, and diplomatic initiatives, or through any combination of these options. Although PR may occur during a noncombatant evacuation operation (NEO), NEO is not a subset of PR (Department of Defense Directive [DODD] 2310.2, *Personnel Recovery*).

SF has a long history of providing support to and conducting unilateral PR operations. Once in the joint special operations area (JSOA), SF can perform its mission unilaterally and with indigenous or surrogate forces or other government agencies (OGAs) to recover isolated personnel. SF units possess the skills, capabilities, and modes of employment to perform PR missions. SF, for example, deploys in small teams trained to operate clandestinely in enemy-held territory or denied areas. These capabilities make SF an effective PR asset in situations where these techniques of rescue and recovery may be preferable, especially in situations when a Special Forces operational detachment A (SFODA) is already present near the PR requirement. SFODAs operating in or near a known evader location can move to this area and then contact, authenticate, support, move, and exfiltrate the evader. SF units, in direct support of a JCSAR operation, may be inserted into hostile territory and travel overland to a predetermined rendezvous point and make contact with the evader. Once contact has been made, the recovery force and the evader move to a location that is within range of friendly air assets for extraction. SF has the unique capability to support theater strategic and operational goals by conducting unconventional warfare (UW). This entails advising, assisting, organizing, training, and equipping indigenous forces and resistance movements in all aspects of UW, to include support to UAR. During the conduct of UAR operations, SFODAs are inserted into enemy territory before hostilities.

EVOLUTION OF SF SUPPORT TO PR

1-1. SF has successfully conducted UAR operations that assisted the movement of isolated personnel from enemy-held, hostile, or sensitive areas to areas under friendly control. A short synopsis of how SF has supported PR is provided below.

WORLD WAR II

1-2. Modern special operations (SO) originated with the vision of Major General (MG) William "Wild Bill" Donovan. During World War II (WW II), MG Donovan was the driving force behind the evolution of U.S. military UW and the creation of the Office of Strategic Services (OSS). The Central Intelligence Agency (CIA) and United States Special Operations Command (USSOCOM) trace their lineage to the OSS. During WW II, renowned OSS SO elements known as "Jedburghs" were employed. Jedburghs were three-man unconventional warfare teams comprised of Americans, British, French, Dutch, and Belgians. They organized resistance groups that conducted UW in Nazi-occupied Europe and established evasion and escape (E&E) nets. Several well-organized and secret escape routes were in operation throughout the war, such as the Comete Line and the Pat O'Leary Line. These lines were designed to assist Frenchmen and Jews fleeing from their German oppressors and Royal Air Force (RAF) and American airmen who had either crash-landed or parachuted to safety after being shot down over Nazi-occupied Europe. The procedures for these lines were the same. Evading aircrew members were passed from link to link in a chain of successive local "helpers" who clothed, fed, and hid them, usually at great personal risk to themselves.

1-3. In July 1942, a recently formed OSS unit was tasked to conduct guerrilla operations in the China-Burma-India theater of operations. This unit was formally designated as Task Force 5405-A, but was better known as Detachment 101 (DET 101). Major (MAJ) Carl Eifler commanded DET 101. The Air Transport Command was responsible for flying the flight route over the Himalayas to deliver supplies to the forces of Chiang Kai-shek. The harsh weather forced down many planes, and the Japanese regularly interdicted the supply operations with their fighters based in Myitkyina (northern Burma). The mission of DET 101 was to conduct guerrilla warfare behind Japanese lines. DET 101 recruited local agents from the Kachin tribe to conduct guerrilla operations against the Japanese in Burma. These recruited agents were trained in tactics, communications, and demolitions and were sent into Northern Burma in small teams to establish operating bases. MAJ Eifler identified PR as a task to be included in their training. While Americans, British, or Anglo-Indians usually led these teams, they usually formed alliances with existing Kachin resistance forces. Forming alliances with existing Kachin resistance forces, DET 101 employed American, British, and Anglo-American American leadership to conduct PR operations. The morale of Allied aircrews flying over the northern Burmese mountains to China improved markedly as OSS teams and agents rescued downed aircrews and brought them back to friendly lines. In all, DET 101 rescued about 400 Allied crewmen. The UAR activities of DET 101 clearly supported the overall war effort, in and beyond Burma, by returning critical pilots, aircrew, and isolated ground troops to the fight.

1-4. The air ground aid section (AGAS) established 39 headquarters scattered from Saigon to the Gobi Desert. The AGAS was responsible for the rescue of 898 airmen. American prisoners of war (PWs) in China, Indochina, and Formosa were released to AGAS officers at the cessation of hostilities. During WW II, there were 130,000 PWs and 78,000 U.S. personnel missing in action (MIA). E&E operations facilitated the recovery of over 47,000 Allied evaders.

KOREAN WAR

1-5. In the years that immediately followed WW II, virtually all paramilitary SOF were disbanded. This conscious decision left the U.S. military without a significant SO capability. What remained of covert operations and UW capability became resident in the newly formed CIA, the successor to the OSS. However, WW II implanted an understanding of the value and potential of SO. Many SO veterans returned to conventional units with valuable operational experience and training. As the Cold War developed, the Army recognized the potential utility UW could contribute to the newly adopted U.S. strategy—containment of Soviet and Chinese communist expansion. During the Korean War, the CIA conducted covert operations with support from the military Services. Three special units were developed and employed to conduct SO behind enemy lines. The United Nations Partisan Infantry, Korea (UNPIK) was a U.S. Army- and CIA-led organization that trained and inserted South Korean nationals behind enemy lines for intelligence collection and sabotage operations. The Combined Command for Reconnaissance Activities, Korea (CCRAK) conducted small-scale sabotage raids along the West Coast of North Korea and its offshore islands. The Joint Advisory Commission, Korea (JACK), a CIA operation in support of the Eighth Army, conducted covert intelligence collection and guerrilla operations from forward bases established on the islands off the east and west coasts of North Korea. The Army provided officers and enlisted men to the CIA to help manage partisan guerrilla activities, to include helping American pilots. The expanded amount of covert operations led to command and control (C2) problems, and the military and CIA UW operations, to include UAR, were never fully synchronized during the Korean War. During the Korean War, there were about 7,000 U.S. PWs, 8,500 U.S. personnel MIA, and 1,000 successful evaders.

COLD WAR ERA

1-6. Recognizing the need for a UW capability to exploit popular disaffection behind the Iron Curtain, the U.S. Army activated the 10th Special Forces Group (Airborne) (SFG[A]) in 1952 and, after training, deployed that unit to Bad Toelz, West Germany. The original missions of the 10th SFG(A) were UW, guerrilla warfare, sabotage, espionage, and E&E. This unit has a long history of planning, exercising, and being prepared to conduct UW, to include establishing E&E networks (now called recovery mechanisms [RMs]) in support of United States European Command (USEUCOM) operational plans to counter the threat of the Warsaw Pact.

VIETNAM WAR

1-7. In the early 1960s, President John F. Kennedy revived interest in UW and other SO in the armed Services. Each Service developed its own counterinsurgency capabilities, and SO units proliferated, fueled by the nation's growing involvement in Southeast Asia and communist-inspired insurgencies in Africa and Latin America. In Vietnam and Laos, USA SF was the key player in training and controlling indigenous and surrogate forces in UW, conducting in-country and cross-border reconnaissance, border surveillance, target strikes, special air missions, and prisoner recovery. In addition to conducting their own operations, the SOF of each Service worked with the CIA on various compartmented programs. An organization that originated and remained closely associated with the CIA was the Military Assistance Command, Vietnam, Studies and Observation Group (MACVSOG). MACVSOG was a joint special operations task force (JSOTF) whose missions included special reconnaissance (SR), direct action (DA), and PR. The Recovery Studies Division, MACVSOG-80, was the joint personnel recovery center. MACVSOG-80 served as the focal point for information and activities related to U.S. and Free World Military Assistance Command missing and captured personnel throughout Southeast Asia. MACVSOG was often called upon to dedicate assets for CSAR operations, which were code named BRIGHT LIGHT. MACVSOG employed BRIGHT LIGHT teams provided by MACVSOG-35 to conduct CSAR missions. As in the Korean conflict, SO, to include PR, were not fully integrated into theater campaign plans. This lack of integration greatly diminished the operational effectiveness of SOF. The Vietnam War saw 591 U.S. POWs, 2,200 U.S. personnel MIA, and more than 4,000 successful evaders.

POST-VIETNAM ERA

1-8. During the post-Vietnam era, the capability of SOF to support PR operations declined. As the U.S. Armed Services focused their attention on the Soviet armor and mechanized threat in Europe, the SF strength of the Army decreased from seven active SF groups to three. The Navy cut the number of sea-air-land teams (SEALs) to 50 percent of wartime levels. The Air Force scheduled all gunships for deactivation, programmed the long-range penetration aircraft for the Reserves, and neglected deep-penetration helicopters. The debilitating effect of this reduction in forces and lack of interagency coordination was tragically demonstrated in the failed attempt by U.S. SOF to rescue the hostages held in Iran in April 1980—Operation EAGLE CLAW. This costly and embarrassing failure revealed serious shortcomings in the ability of the U.S military to plan and conduct a complex, high-risk special liberation mission.

1-9. In 1985, the Services promulgated that unconventional recovery of personnel from enemy-controlled or politically sensitive territory would be the responsibility of the theater Special Operations Command (SOC). The principal missions of SF include UW. UAR is a sub-mission of UW. In the conduct of UAR, SF teams are directed to contact, authenticate, support, move, and return personnel from enemy-held or hostile areas to friendly control. The general concept entails using SOF in PR operations to link up the survivor with a recovery force as soon as possible and move the individual to an area under friendly control. The Services conceived that SF could provide assisted recovery to support designated personnel when the threat exceeded conventional recovery capabilities.

DESERT SHIELD AND DESERT STORM

1-10. During Operations DESERT SHIELD and DESERT STORM, SF units provided UW support to the Kuwaiti resistance. During post-DESERT STORM operations, SF has developed a fully integrated UAR capability in support of the combatant commanders' PR operations.

OPERATION ENDURING FREEDOM

1-11. Operation ENDURING FREEDOM has seen the resurrection of UW. This operation also resurrected the vital role of SF in training and leading indigenous forces in UW, to include conducting UAR.

DEPARTMENT OF DEFENSE POLICY

1-12. One of the highest priorities of the Department of Defense (DOD) is preserving the lives and well-being of U.S. military, DOD civilians, and contract service employees in danger of being isolated, beleaguered, detained, captured, or having to evade while participating in a U.S.-sponsored activity or mission. The DOD has a moral obligation to—

- Return its personnel to friendly control.
- Protect its personnel and every living creature.
- Maintain morale.
- Deny the enemy a potential source of intelligence.
- Deny the use of captured personnel in propaganda programs designed to influence our national interest and military strategy.
- Prevent exploitation of its personnel by adversaries.
- Reduce the potential for use of captured personnel as leverage against the United States.

1-13. The DOD has primary responsibility for recovering U.S. military, DOD civilian, and contract service employees deployed outside the United States and its territories. When requested, and when directed by the President and/or the Secretary of Defense (SecDef), the DOD shall provide PR support to other governments, agencies, and organizations, IAW applicable laws, regulations, and memorandums of agreement or understanding.

USSOCOM POLICY

1-14. USSOCOM Directive 525-21, *Personnel Recovery*, establishes the basis for SOF PR doctrine. The Special Operations Center, Operating Plans—Train, Doctrine, Education Division (SOOP-PT) is the PR office of primary responsibility (OPR) within the USSOCOM staff. SOF are responsible for the recovery of SO personnel, usually through UAR and, when directed by a theater geographic combatant commander, JCSAR.

CSAR AND SELF-RECOVERY

1-15. USSOCOM maintains a Service-like responsibility to perform CSAR in support of its own operations, consistent with capabilities and assigned functions and IAW the requirements of the supported commander. SOF must maintain an inherent and organic capability to conduct self-recovery within its core mission force structure. Self-recovery and emergency exfiltration operations must be an inherent part of every SO mission organic to its core mission function.

JCSAR

1-16. SOF may be directed to support JCSAR missions. However, JCSAR taskings may be at the expense of core SOF mission readiness and/or capabilities. SOF is highly suited for JCSAR because of—

- Its normal operational area or environment.
- Its unique ability to penetrate hostile defense systems and conduct air, ground, or sea operations deep within hostile or denied territory at night or in adverse weather.

1-17. The Commander, Air Force Special Operations Command (AFSOC/CC) is designated as the USSOCOM proponent for CSAR or JCSAR matters. AFSOC Instruction 10-3001, *Personnel Recovery*, provides the framework for PR as it pertains to Air Force Special Operations Forces (AFSOF).

UAR

1-18. SOF develop and execute UAR across the spectrum of conflict in all environments. UAR, a subset of NAR, may be conducted unilaterally, with indigenous assets, or with OGAs. SOF may develop a unilateral UAR capability as an organic force protection measure. The Commanding General, USASOC, is designated as proponent for USSOCOM in UAR matters.

SF PR

1-19. SF units are responsible for self-recovery in support of their own operations, consistent with organic capabilities and assigned functions and IAW the requirements of the supported commander. SF units must make recovery planning an inherent part of every mission and include recovery and emergency exfiltration operations.

PHILOSOPHY BEHIND PR OPERATIONS

1-20. Successful PR requires effective training and extensive preparation and planning. PR planners must identify, schedule, and commit the necessary resources as soon as possible. However, evaders must be prepared for extended evasion should recovery operations be delayed.

FIVE TASKS OF PR

1-21. The DOD PR system exists to ensure a complete and coordinated effort to recover DOD soldiers, DOD civilians, and DOD contractor personnel who become captured, detained, or otherwise isolated from U.S. control. As shown in Figure 1-1 page 1-7, PR consists of activities to prepare for and conduct operations to report, locate, support, recover, and return or repatriate personnel who have become isolated from friendly forces.

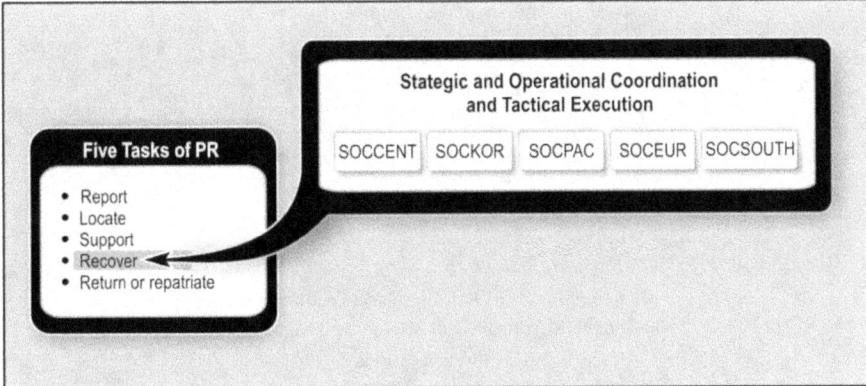

Figure 1-1. PR Tasks

REPORT

1-22. *Report* is the accurate and timely realization, confirmation, and dissemination that an individual has become isolated, captured, or detained. The isolated individual or another element or individual that observed the isolating incident can forward the initial report. Reporting may include immediate secure voice or data reporting of all incidents by—

- Using required PR report formats
- Preparing additional reports as directed by the theater JSRC and the joint force commander (JFC).
- Recording all information received about a given incident.

LOCATE

1-23. *Locate* is all actions taken to pinpoint the exact geographic position of isolated personnel and to pass the resulting information to the appropriate organizations for coordination and action. Efforts to locate isolated personnel begin with their known location. Time, effort, and lives can be lost if the exact position of isolated personnel is not accurately verified in a reasonable time. All available assets should be considered to locate the terminal objective area. If the location is not readily discernible, the JFC issues guidance on the search methods. Search methods include, but are not limited to, electronic detection, visual systems, sensor systems, and physical ground searches.

SUPPORT

1-24. *Support* is planned efforts that sustain mentally, physically, and emotionally the isolated personnel until recovery can be affected. Support should consist of systems that are established, validated, and disseminated to

RECOVER

all potential isolated personnel before the isolating incident. Comprehensive knowledge of DOD efforts to recover and repatriate isolated personnel and efforts to provide family support will reduce mental and emotional stress for the isolated personnel and their family members.

1-25. *Recover* is the return of evaders to friendly control, either with or without assistance, as the result of planning, operations, and individual actions on the part of recovery planners, recovery forces, and the evaders themselves. Determining recovery force options involves many critical factors. Factors to consider are:

- Component go/no-go criteria.
- Threat assessment.
- National assets.
- Current operations. (Can missions be diverted or rerolled?)
- Command, control, and communications.
- Airspace and ground operations deconfliction.
- Psychological operations (PSYOP).
- Civil affairs (CA).
- Preplanned CSAR responses.
- Isolated personnel consideration.
- Authentication of isolated personnel.
- Department of Defense (DD) Form 1833, *Isolated Personnel Report*.
- Theater code words (special instructions [SPINS]).
- Visual and electronic signals (SPINS).

REPATRIATE

1-26. *Repatriate* is all efforts to smoothly return the isolated personnel to their previous life or return their remains to their next of kin. Repatriation encompasses the entire process from gathering lessons learned from a PR event to supporting a smooth reintegration of the isolated personnel into their units and families. Repatriation is also the process that covers medical treatment; intelligence debriefings; survival, evasion, resistance, escape (SERE) debriefings; psychological assistance; and next-of-kin support. The repatriation process is designed to maximize the accuracy of information the returnee provides during debriefings, while ensuring optimal personal support for decompression, medical treatment, and reintegration into his family and unit.

RECOVERY CONSIDERATIONS

1-27. Recovery considerations that may have an effect on any recovery operation are—

- *Availability of resources.* PR planners must identify requirements in support of component operations and develop potential COAs to meet the commander's intent. To increase the chances for successful recovery, planners should identify well in advance the resources that will be required for a particular operation and schedule them for employment. Advance planning ensures these resources are available when required.

- *Capabilities and limitations.* PR planners must have in-depth knowledge of the capabilities and limitations of each type of resource used for recovery operations. Such in-depth knowledge enhances the chances for successful recovery, because recovery planners can—
    - Quickly, properly, efficiently, and effectively develop required task organization.
    - Quickly develop alternative COAs and options in response to changing recovery scenarios.
- *Task organization.* The nature of any given conflict (specific purpose, intensity, or duration) that might leave individuals isolated in hostile territory will also dictate the capabilities and resources required to accomplish the recovery mission. Consequently, task-organizing the recovery force depends upon varying resource capabilities and requirements. For instance, even the least sophisticated weaponry employed by enemy forces can be lethal to unescorted CSAR recovery vehicles. Several factors may require the formation of a cohesive CSAR effort consisting of similar or dissimilar aircraft and forces. Among these factors are—
    - The concentration of enemy weapons and troops.
    - The enemy's degree of integration with other defensive systems or command, control, communications, computers, and intelligence (C4I) networks.
    - The accuracy and timeliness of friendly intelligence data.
    - The number of personnel requiring recovery.
    - The location and physical condition of the isolated personnel.

    Therefore, two or more resources may have to be task-organized to recover an isolated person. The assembly of two or more assets to support a single CSAR effort is referred to as a combat search and rescue task force (CSARTF). A CSARTF requires thorough premission and real-time planning and coordination with participating elements.

- *Recovery criteria.* The decision maker needs certain information requirements (IRs) and priority intelligence requirements (PIR), when available, to simplify the recovery effort. The IR and PIR might include the following:
    - *Location and physical condition of the evader.* Accurate information concerning the location and physical condition of the evader is crucial before the recovery mission is executed. With this knowledge, planners can more readily determine the optimum force composition, identify any requirements for special equipment and personnel, and plan recovery force ingress and egress routes to minimize contact with hostile forces.

- *Access.* A critical factor in the selection of any recovery site is air, land, or sea access to the site. PR planners must be familiar with the capabilities and limitations of recovery platforms. PR planners must also consider the spatial relationships of the site with respect to bordering friendly, enemy, and neutral territories. For example, authorization for overflight of a bordering territory may be imperative to the viability of a particular PR COA.
- *Time.* Recovery assets become operational at different times. Dedicated CSAR assets are generally operational earlier than those without a specified PR mission. UAR assets are normally in place well before hostilities although their reaction time to a PR event can be slowed by communications and security considerations. To permit interface with available recovery forces, evasion planners must ensure that all potential evaders have access to appropriate contact and communications procedures. Thorough prior planning permits operations personnel to predict when recovery assets are available to them. Planners should also ensure that potential evaders are updated in a timely manner as changes occur in the operational recovery environment.
- *Movement.* Using dedicated CSAR forces is the preferred mode of recovery whenever evaders are within range and the military situation permits. Isolated individuals will not always be within the range of conventional recovery assets or in areas where recovery resources are operational. The political situation may also impact on the method of PR operations executed. For instance, in conducting military operations other than war (MOOTW) where the presence of U.S. forces is denied, disavowed, or politically sensitive, recovery operations may have to be conducted in a covert or clandestine manner. Therefore, isolated personnel may be required to hide or to move to a predetermined recovery area to be contacted by recovery forces. Depending on the access or freedom of movement the recovery element has within its area of operation, it could be required to move the isolated personnel several times by using crossover procedures between various elements.
- *Capacity.* Recovery forces and assets are based or pre-positioned in locations where they can most effectively facilitate recovery operations. Since individuals may become isolated in unexpected areas and numbers, in some cases the capacity of available recovery forces may be inadequate for the number of evaders. PR planners must be prepared to reallocate recovery assets to compensate for this imbalance or to establish priorities to determine which category of evaders will be recovered by a limited capacity of recovery resources. If the estimated capacity of recovery assets appears inadequate for a given number of evaders, planners must ensure that potential evaders are aware of the situation and that they are trained and equipped for extended survival and evasion.
- *Risk assessment.* The benefit to be gained from a recovery operation must be weighed against the anticipated risk of execution. Recovery operations should not unduly risk isolating additional

combat personnel, preclude the execution of higher priority missions, routinely expose certain unique assets to extremely high risk, divert critically needed forces from ongoing operations, or allow the military situation to deteriorate.

PR SPECTRUM

1-28. Figure 1-2 depicts the PR spectrum as viewed by DOD. Within the PR spectrum, SAR is the DOD conducting search and rescue for DOD personnel in a permissive environment as defined by the Missing Person Act. There are large numbers of capable SAR resources with relatively easy access worldwide. PR operations become more complex when there is a threat. Once a threat is present, generic SAR is no longer an option and other applications of the PR spectrum must be pursued. As the threat increases, risk increases, and PR planners must adjust the planning time line to consciously plan and execute recovery missions.

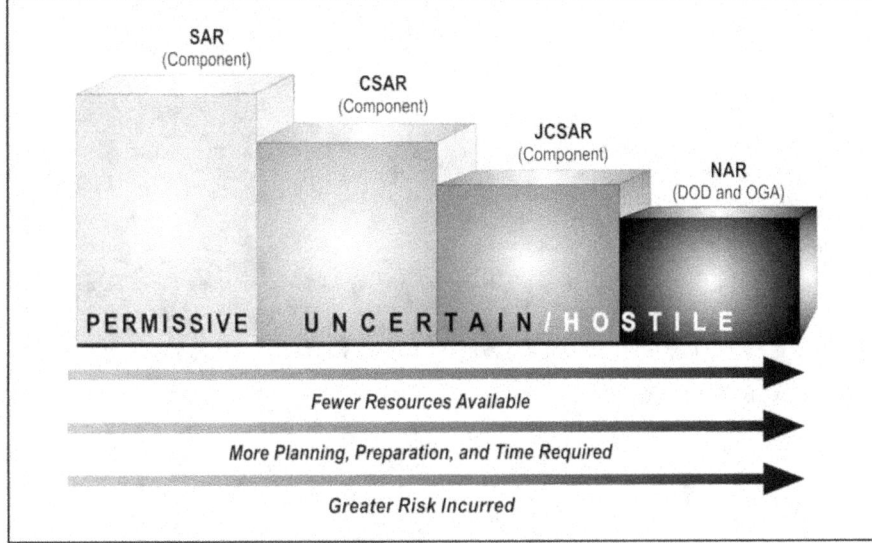

Figure 1-2. PR Spectrum

1-29. Each Service component has the responsibility to plan for the recovery of its own forces; that is, CSAR. CSAR can be conducted in uncertain to hostile environments. The availability of resources and their capabilities and limitations, location and physical condition of the evader, access, time, movement, and risk management play an important part in the decision-making process that determines whether the JSRC selects a CSAR, JCSAR, or a NAR option for the recovery effort. CSAR forces of a single component normally conduct rescue operations with dedicated rotary-wing aircraft, specially trained aircrews, and support personnel in response to a PR tasking.

1-30. If the PR tasking exceeds the capability of a single component to perform the recovery, a JCSAR mission may be conducted. A joint CSARTF is a mutually supporting package of assets tailored to meet a specific CSAR requirement. CSARTF resources provide a variety of services. These services include command, control, and communications; location and authentication of isolated personnel; protection of isolated personnel and task force elements from air and ground attacks; navigation assistance; armed escort, combat air patrol; and aerial refueling support.

1-31. When conventional recovery forces are not present in an uncertain or hostile environment or their presence is infeasible or unacceptable, the use of NAR forces may be the only viable option. NAR or UAR options are advantageous in areas where the enemy air or ground threat or other factors such as weather and terrain prevents conventional recovery or when a clandestine or precisely timed recovery operation is required because of threat levels or political sensitivities. SF UAR resources are a valuable commodity and must be established and maintained in advance of their potential need. UAR operations require more planning and preparation, and the execution time line may be measured in days instead of hours.

OPTIONS AND METHODS OF PR

1-32. Recovery of evaders may occur because of diplomatic negotiations, planned military operations, or civil actions by private organizations. Figure 1-3, page 1-13, depicts the current PR options and methods.

CIVIL OPTION

1-33. Civil options are the sum of civil efforts, working alone or in concert with diplomatic and military efforts, to recover isolated personnel through NGOs. The release of Navy pilot Lieutenant (LT) Robert Goodman from Syria in 1984 is an example of a nonofficial civil option.

DIPLOMATIC OPTION

1-34. A diplomatic option is the sum of diplomatic efforts, working alone or with military and civil efforts, to recover isolated personnel through diplomatic channels. The Department of State (DOS) is chartered as the lead agency for diplomatic actions and negotiations to effect the release of isolated personnel back to friendly control. A recent example of the diplomatic recovery option is the return of the EP-3 crew from China because of intensive diplomatic efforts between U.S. and Chinese officials.

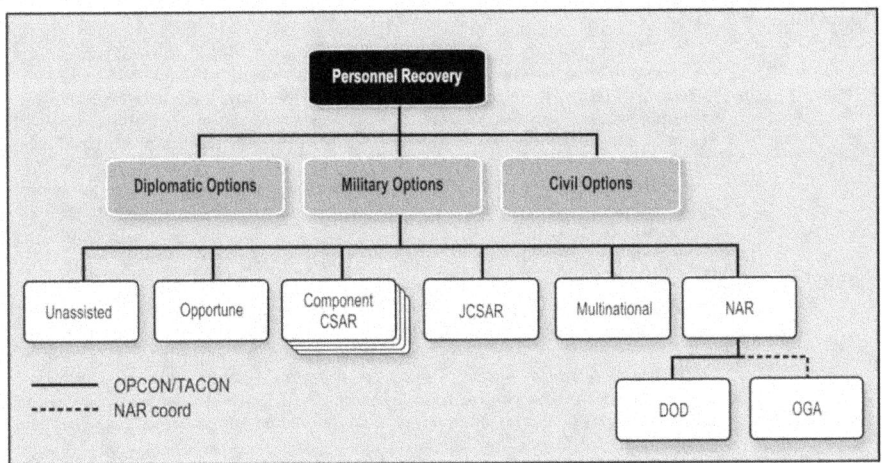

Figure 1-3. PR Options and Methods

MILITARY OPTION

1-35. Military options include the efforts of isolated personnel to make their own way back to friendly control as a result of their detailed evasion plan of action (EPA) planning and recovery operations conducted by CSAR and NAR or UAR forces. Military efforts may include working with diplomatic and civil efforts to perform PR. The recoveries of "Vega 31" and "Hammer 34" by a joint CSARTF in Serbia are good examples of individuals recovered via the military recovery option. The PR methods of the military option are discussed in the following paragraphs.

Unassisted

1-36. The unassisted PR method is PR performed through the independent efforts of missing or isolated personnel by conducting successful evasion to friendly forces. Evaders may simply hide and survive in a single location while waiting to be overrun by friendly forces or evade back to friendly or neutral territory. Unassisted recovery is normally a contingency to be used if recovery forces cannot gain access to the isolated individual. Successful unassisted recovery depends largely on the evader's will and ability, personal background, physical condition, and previous SERE training. Evaders may need to travel long distances over unfamiliar terrain, suffering long periods of hunger, thirst, and exposure. The primary concern of evaders facing these situations is to reach a location where recovery can be effected. Air-delivered or pre-positioned supplies (caches) of evasion and survival equipment can greatly improve the evader's potential for success. Because extended unassisted recovery is always a possibility, every EPA must address it as a contingency.

Opportune

1-37. The opportune method is PR performed by military forces, indigenous persons, or others who are not specifically trained and dedicated to PR but are coincidentally nearby and available to perform the recovery mission. Component air, ground, or naval forces, although not specifically trained in combat recovery, may be in the area of an isolated individual and could be tasked to recover the evader. In all cases, however, the recovery effort should be coordinated with the JSRC, which continually monitors all ongoing and planned recovery operations. Use of opportune resources would most likely be the result of chance.

Component CSAR

1-38. The component method is a specific task performed by component rescue forces to effect the recovery of isolated personnel during major theater war or MOOTW. CSAR is performed by dedicated, appropriately equipped military forces of a single Service or functional component using organic resources trained in the employment of PR tactics, techniques, and procedures (TTP). The component RCC informs and coordinates with the J3RC. The JSRC monitors the situation.

JCSAR

1-39. When the requirements of the recovery mission exceed the capabilities of a single component, the JFC normally establishes a CSARTF to conduct JCSAR (Figure 1-4, page 1-16). Coordinated by and under the direction of the JSRC, JCSAR operations require the effort of two or more components of the joint forces. The CSARTF consists of all forces committed to a specific CSAR operation to search for, locate, identify, and recover isolated personnel during wartime or contingency operations. A CSARTF is a mutually supporting package of assets tailored to meet a specific CSAR requirement. A CSARTF may consist of the following elements:

- *Airborne mission commander.* An airborne mission commander (AMC) may be designated by component RCCs or higher authority to coordinate the efforts of several assets. The AMC serves as an extension of the RCC, and if required, designates the on-scene commander (OSC). The AMC coordinates and controls the flying mission for forces designated to support a specific CSAR operation. The AMC's responsibilities include coordinating the CSARTF radio net, managing the flow of aircraft to and from the objective area, coordinating for additional CSARTF support, and monitoring the tactical air and ground situation in and around the objective area and CSARTF.

- *On-scene commander.* The OSC is the individual designated to control rescue efforts at the rescue site. The rescue escort (RESCORT) flight lead [Sandy 1] is most often designated OSC. However, a wingman or rescue combat air patrol (RESCAP) aircrew member may function as the OSC until the arrival of either the AMC or RESCORT [Sandy 1] aircraft. All CSARTF participants must clearly understand that a transfer of the OSC role has occurred. The OSC helps to ensure effective asset management in the often chaotic and hostile objective area.

- *Recovery vehicles.* Typically, a primary and a secondary recovery vehicle are flown to the objective area to make the pickup. This formation provides a backup mission aircraft and offers mutual support should the primary recovery vehicle have problems. The secondary recovery vehicle should be prepared to assume lead responsibilities and accomplish the recovery should the lead aircraft abort the mission or be unable to perform primary recovery responsibilities.

- *RESCORT.* RESCORT aircraft protect rescue assets from surface threats during ingress and egress and while in the objective area. RESCORT aircraft also provide navigational assistance, armed escort, and assistance in locating and authenticating isolated personnel.

- *RESCAP.* RESCAP aircraft are air superiority assets assigned to protect the CSARTF from airborne threats while en route to and returning from the objective. Once established, RESCAP aircraft also maintain air patrol over the objective area. RESCAP aircraft may also assist RESCORT aircraft in locating and authenticating isolated personnel.

- *Special tactics teams.* Special tactics teams (STTs) are ground combat forces assigned to AFSOC. They are composed of combat controllers and pararescuemen (PJ) specifically organized, trained, and equipped to facilitate and expedite the use of aviation assets and to provide CSAR expertise when requested through appropriate channels. STTs may be augmented by security elements comprised of other ground forces such as SF or U.S. Army Rangers. STTs—

    - Provide PR and CSAR planning expertise.
    - Facilitate contact, authentication, and mechanical extrication.
    - Provide medical treatment at the paramedic level.
    - Facilitate movement and exfiltration during the recovery of personnel.
    - Provide equipment not accessible to conventional CSAR resources.

- *Fixed-wing tanker assets.* Fixed-wing tanker-capable rescue assets are a key element of CSAR operations. These assets play a critical role in extending the operational range of air-refuelable helicopters.

- *Other forces.* CSARTF operations may occur in day, night, or adverse weather conditions. Operations may also require onboard countermeasures and force packaging to defeat threats. Gunships, suppression of enemy air defenses (SEAD) aircraft, and electronic warfare aircraft may assist in the recovery effort.

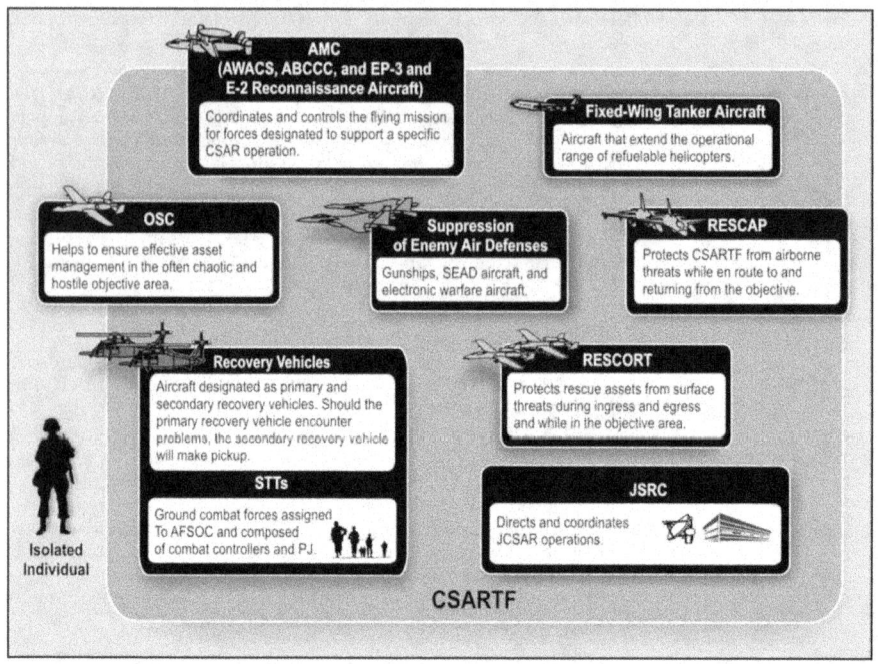

Figure 1-4. CSARTF for JCSAR

Multinational

1-40. The multinational method is PR operations performed by a dedicated, appropriately equipped multinational force trained in the employment of PR TTP. The multinational method is conducted with U.S. alliance or coalition partners.

NAR

1-41. The NAR method is PR performed by dedicated or designated forces trained in the employment of compartmented TTPs. It is DOD policy to complement its PR capabilities with NAR to recover isolated personnel in those instances when the use of conventional recovery forces in enemy-held or hostile areas is infeasible, unacceptable, or nonexistent. NAR operations may be covert or clandestine. NAR requires preconflict deliberate planning, training, and support to reduce risk by developing assets and credible capabilities. These PR capabilities are especially advantageous in areas where enemy air or ground threat prevents conventional recovery operations or when a clandestine or precisely timed recovery operation is required because of threat levels or political sensitivities. As defined in Department of

Defense Instruction (DODI) 2310.6, *Non-Conventional Assisted Recovery in the Department of Defense*, NAR is all forms of PR conducted by an entity, group of entities, or organizations that are trained and directed to contact, authenticate, support, move, and exfiltrate U.S. military and other designated personnel from enemy-held or hostile areas to friendly control through established infrastructure or procedures. NAR includes UAR. Per United States Code (USC) Title 10, Section 1501, *System for Accounting for Missing Persons*, DODI 2310.6, and USSOCOM Directive 525-21, UAR is NAR conducted by SOF, specifically U.S. Army SF and U.S. Navy SEALs. UAR is a subset and integral component of UW. SOF (SF and SEALs), OGAs, and multinational forces may provide NAR assets to conduct NAR operations. NAR entails the employment of recovery teams (RTs) and RMs. The distinct differences of NAR from conventional recovery include the following:

- Potential use of indigenous or surrogate assets.
- General reliance on a robust infrastructure developed by trained NAR forces.
- Prestrike planning and deployment of forces.
- Specialized training, equipment, and capabilities.
- Operational techniques, environment, and unorthodox approach.
- Higher degree of political risk.
- Dependence on detailed operational intelligence and operations security (OPSEC).
- Potential to operate independent from friendly support.
- Longer planning timelines to develop, train, and equip.
- Limited opportunity and resources.

1-42. NAR is focused on the recovery aspect of the DOD PR system and involves the following five specific tasks as shown in Figure 1-5, page 1-18, and described below:

- *Contact.* Contact consists of predesignated actions taken by isolated personnel and recovery forces to facilitate linkup between the two parties in hostile territory to effect the eventual return of those isolated personnel to friendly control.
- *Authenticate.* Authentication is the process whereby the identity of an isolated individual is confirmed.
- *Support.* Support is all of the actions taken to safeguard an isolated person, monitor his physical and mental well-being, and provide him with shelter, food, and clothing.
- *Move.* Movement may entail transporting isolated personnel from a contact point to, or between, one or more elements of an infrastructure (safe site or holding area, crossover or hand-over site, and NAR forces) and, ultimately, to a final exfiltration site.
- *Exfiltrate to friendly control.* Exfiltrate to friendly control is the final transfer of isolated personnel by NAR forces to definitive U.S. Government control.

FM 3-05.231

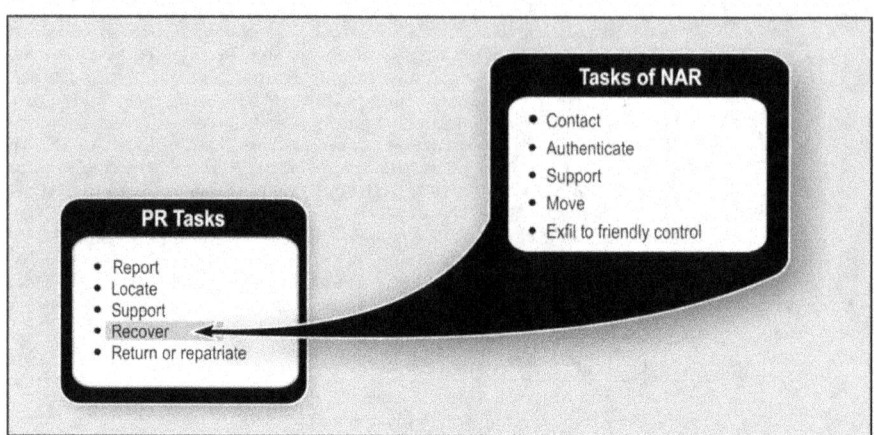

Figure 1-5. NAR Tasks

Chapter 2

Command, Control, Coordination, and Mission Management

JFCs must develop a comprehensive, fully integrated PR program. The joint force land component commander (JFLCC), joint force marine component commander (JFMCC), joint force air component commander (JFACC) and joint force special operations component commander (JFSOCC) must perform CSAR in support of their own operations. The JFSOCC establishes a SOC RCC, or functional equivalent, as the focal point for interface with the JSRC to coordinate all SOF CSAR activities. The JFSOCC is also responsible to the JFC for all NAR operations in support of the JFC's comprehensive PR program.

COMMAND RELATIONSHIPS

2-1. JFCs, to include geographic combatant commanders and JTF commanders, have a directed command responsibility to develop a comprehensive, fully integrated PR program and to incorporate CSAR and NAR capabilities into their AORs or joint operations areas (JOAs). The JFC generally exercises his command responsibility for the planning and conduct of PR operations through the theater component commander possessing the most forces; command, control, communications, computers, intelligence, surveillance, and reconnaissance (C4ISR); and battlespace awareness to manage PR across the theater. Normally, the JFC uses the JFACC, because the air component usually has the most recovery and support platforms and manages the air war. To coordinate and integrate planning and operations capabilities for PR within the AOR, the JFC is tasked to establish a JSRC or functional equivalent. The JFC may choose to establish the JSRC director as a member of his special staff to centrally manage PR operations, or he may elect to task a theater component—normally the JFACC—to perform those duties.

2-2. The JFLCC, JFMCC, JFACC, and JFSOCC are tasked with command responsibility to perform CSAR in support of their own operations. These theater component commanders ensure their command performs CSAR consistent with component capabilities and assigned functions and in accordance with (IAW) the requirements of the supported combatant commander. To coordinate all component CSAR activities and integrate CSAR capabilities within the supported commander's PR program, each theater component commander is tasked to establish an RCC or functional equivalent. Each theater component commander has operational control (OPCON) and tactical control (TACON) of his RCC. Furthermore, each RCC is directed to establish expeditious lines of communication vertically with the

FM 3-05.231

JSRC and laterally with the other theater component RCCs. Figure 2-1 shows the typical theater PR architecture.

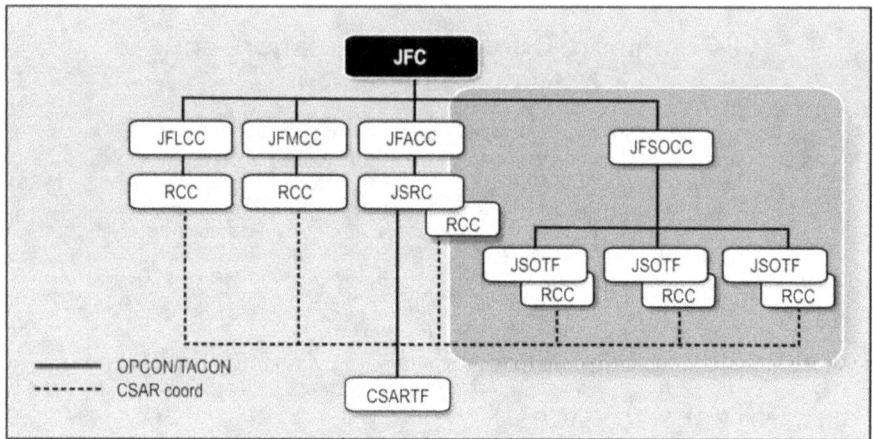

Figure 2-1. Typical Theater PR Architecture

2-3. The JFSOCC, operating either as a theater SOC or as a JSOTF, has command responsibility to perform CSAR of his forces consistent with capabilities and assigned functions and IAW requirements of the supported combatant commander. The JFSOCC establishes his SOC RCC, or functional equivalent, as the focal point for interface with the JSRC to coordinate all SOF CSAR activities. The JFSOCC may choose to task the air operations section within its joint operations center (JOC) to coordinate all SOF CSAR activities (Figure 2-2, page 2-3). The joint special operations aviation component (JSOAC) is manned, trained, and equipped for CSAR. The JSOAC is the AFSOC proponent for SOF CSAR TTP and communications infrastructure. As such, the JFSOCC may elect to task the JSOAC to perform the SOC RCC functions (Figure 2-3, page 2-3).

2-4. The JFSOCC integrates and deconflicts the deliberate planning of SOF operations via the joint special operations liaison element (JSOLE), which is generally located in the joint combined air operations center (JAOC) or combined air operations center (CAOC). The JSOLE conducts synchronization of SOF plans within the standard air tasking order (ATO) production schedule. Because of the time sensitivity of a dynamic PR operation, the SOC RCC coordinates directly with the JSRC. The JFSOCC may relinquish TACON of SOF to the JFACC when necessary to conduct specific JCSAR operations in support of a CSARTF.

2-5. Not only is the JFSOCC responsible for CSAR, he is responsible to the JFC for all NAR operations in support of the JFC's comprehensive PR program. The JFSOCC retains command authority of all SOF UAR forces in the theater. OGAs in support of NAR normally retain C2 of their respective forces. To integrate, coordinate, and synchronize compartmented NAR support to the JFC's PR plans and procedures, the JFSOCC, on order, establishes a UARCC with clear and expeditious lines of communication with the JSRC.

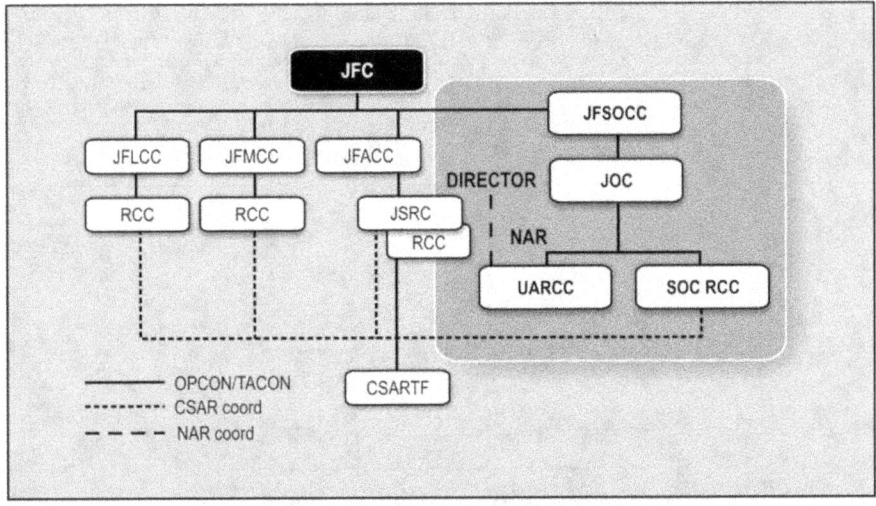

Figure 2-2. SOF PR Architecture, JOC Air Operations Center

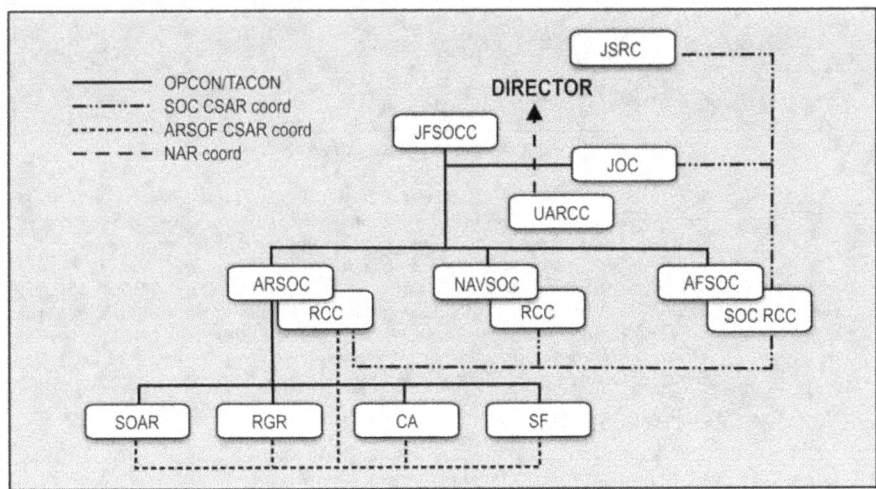

Figure 2-3. SOF PR Architecture, JSOAC

2-6. Similarly, the Commander, Army Special Operations Task Force (ARSOTF), has a command responsibility for CSAR of his forces consistent with capabilities and assigned function and IAW requirements of the supported combatant commander. To coordinate component CSAR, he should establish the functional equivalent of an RCC to interface with the existing PR architecture. The ARSOTF commander may task those functions to the air operations section within his operations center (OPCEN) or, given the number of recovery platforms and communications infrastructure, his Army SO aviation component. The ARSOTF commander may relinquish TACON of ARSOF to the JFSOCC or JFACC, as appropriate, to conduct specific JCSAR operations in support of a CSARTF (Figure 2-4).

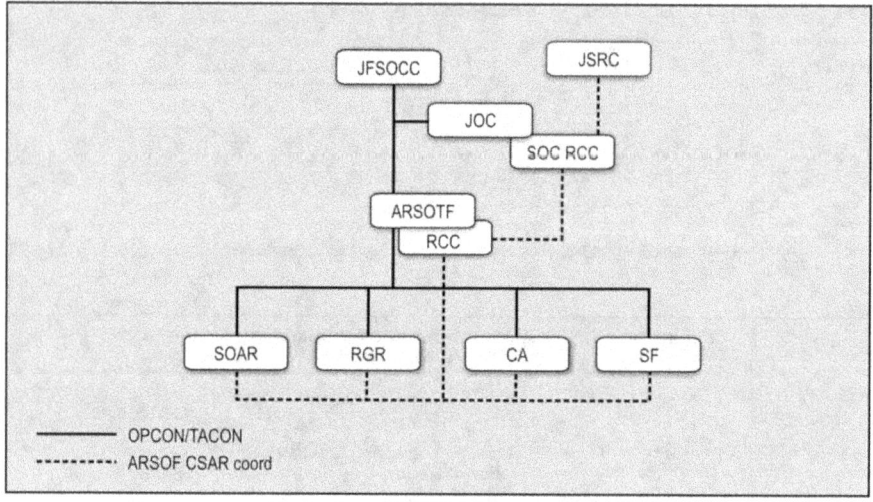

Figure 2-4. ARSOTF PR Architecture

2-7. The Special Forces operational base (SFOB) commander, has command responsibility for CSAR of his forces consistent with capabilities and assigned functions and IAW requirements of the supported combatant commander. SFOB commanders are not required to establish an RCC; however, they should task PR responsibilities and activities to dedicated PR coordinators located within the OPCEN. The SFOB commander coordinates, via the OPCEN, with the joint force special operations component (JFSOC) for PR support. Furthermore, the SFOB commander must be prepared to accept forces to form an ARSOTF or JSOTF.

LAUNCH-AND-EXECUTE AUTHORITY

2-8. Command relationships establishing launch-and-execute authority must be clearly articulated before the commencement of military operations in general, but more specifically, before PR operations begin. The JFC may retain launch-and-execute authority or delegate that authority to the JFACC for CSAR events. Similarly, for NAR events, the JFC may retain launch-and-execute authority or choose to delegate that authority to the JFACC or JFSOCC.

COORDINATION OF RECOVERY PERSONNEL

2-9. The theater RCCs include the JSRC, theater component RCCs, and the UARCC. The JSRC plans, coordinates, and executes JCSAR operations. Theater component commanders establish component RCCs. These component RCCs support CSAR and coordinate vertically with the JSRC and laterally with the other theater component RCCs. The UARCC integrates and coordinates all theater NAR capabilities in support of the JFC's PR program. The UARCC receives input from the JSRC and theater component RCCs. However, because of the compartmented nature of NAR, the UARCC communicates concepts of operation (CONOPS) and operational information exclusively with the JSRC director. Although not formally tasked to establish an RCC, SOF component commanders should be prepared, according to their inherent capabilities, to plan, coordinate, and conduct PR in support of their own operations and provide mutual PR support, as required. Therefore, ARSOTF and SFOB OPCEN staffs must be prepared to interface and coordinate with the SOC RCC in all aspects of PR—especially if the SFOB is designated a JSOTF.

JSRC

2-10. The JSRC is the primary theater PR facility. Specially trained for PR operations, the JSRC is suitably staffed by supervisory personnel. The JSRC is also equipped for planning, coordinating, and executing joint PR operations. Theater component commanders should be prepared to provide augmentation (communications specialists, intelligence specialists, SERE personnel, and so on) and support to the JSRC. Augmentation and support may include United States Coast Guard (USCG) representation, where appropriate. In a coalition environment, the JSRC may have a multinational staff. The JSRC normally develops integrated PR CONOPS, annexes and appendixes to support operation plans (OPLANs), operation plans in concept format (CONPLANs), and peacetime operations. A standing JSRC may also coordinate PR training and exercises to provide a trained joint staff element for combat operations. Other typical CSAR-related JSRC responsibilities and functions in peacetime and combat are discussed below.

Peacetime Operations

2-11. During peacetime operations, the JSRC—

- Assists in the development of PR appendixes to Annex C (Operations) to OPLANs, CONPLANs, and operation orders (OPORDs). Ensures the PR appendixes address the repatriation process and are linked to related appendixes for casualty affairs, medical support, repatriation, and mortuary affairs.
- Develops an integrated PR CONOPS to support operations.
- Develops joint force PR standing operating procedures (SOPs).

- Conducts or provides on-the-job training and informal training for JSRC personnel and component RCC augmentation personnel.
- Develops PR communications plans.
- Establishes reporting requirements for the JSRC and component RCCs.
- Coordinates and deconflicts component PR plans and reviews them for supportability.
- Organizes and conducts PR mission training exercises for the joint force.
- Develops a plan to transition from peacetime to combat operations by—
  - Developing augmentation personnel requirements.
  - Establishing additional communications support requirements.
  - Establishing dedicated intelligence support requirements, to include joint intelligence center (JIC) and joint intelligence support element (JISE) support requirements.

Combat Operations

2-12. During combat operations, the JSRC—
- Develops a joint force PR threat decision matrix tailored to the current threat analysis.
- Develops and disseminates CSAR SPINS, to be included in ATOs, which include the theater PR guidance, concepts, and specific procedures for all high-risk combatants.
- Coordinates themes that support PR with the joint force PSYOP officer. The themes are incorporated into the JFC's PSYOP campaign plan to favorably influence the local population regarding PR efforts.
- Alerts theater component RCCs of the known or suspected locations of isolated personnel by disseminating search and rescue incident reports (SARIRs) and search and rescue situation reports (SARSITs).
- Coordinates with national, theater JIC, host nation, and component intelligence resources to gather information relating to the location and status of isolated personnel and the threat that may affect their successful recovery.
- Coordinates and deconflicts joint force and multinational CSAR support operations. Has tasking authority to coordinate planning for JCSAR operations. Has the authority to coordinate support to the UARCC for NAR operations.
- Monitors all PR incidents executed by the component RCCs.
- Maintains a database and file on each isolated individual until recovery is complete. Forwards, but does **not** destroy, all files and the database to the Joint Personnel Recovery Agency (JPRA) once the recovery mission is complete and the JFC no longer has a requirement to maintain the files.

RCC

2-13. Joint force component commanders (JFCCs) are responsible for planning and conducting CSAR operations in support of their own operations while executing the JFC's campaign and OPLANs. The RCC is the primary component search and rescue facility. The RCC is suitably staffed by trained supervisory personnel and equipped for coordinating and controlling component PR operations. The component RCC coordinates CSAR activities with the JSRC and other component RCCs, as appropriate. If a component commander does not establish an RCC, another component staff organization, usually the operations section (G-3/J-3), coordinates those CSAR activities and assumes the responsibilities normally assigned to the RCC. For example, the JFSOCC normally designates its JOC to coordinate CSAR operations that use SOF component forces. Other typical RCC responsibilities and functions include the following:

- Develop and review PR appendixes to Annex C (Operations) for component supporting plans, CONPLANs, and OPORDs.
- Extract planning factors from the theater PR CONOPS and provide them to subordinate units to assist in the development of supportable and feasible EPAs.
- Coordinate and develop unit and personal EPAs, as required. (Appendix A provides an example of an EPA format, Appendix B provides an example format for the FOB EPA guidance format, and Appendix C discusses EPA planning.)
- Maintain staff awareness for preparing and maintaining isolated personnel reports (ISOPREPs) (Figure 2-5, pages 2-8 and 2-9).
- Coordinate for component augmentation personnel to the JSRC.
- Develop specific component RCC procedures.
- Notify and coordinate with the JSRC when conducting unilateral CSAR missions.
- Coordinate with the JSRC for CSAR support; for example, RESCAP, RESCORT, SEAD, and so on, provided to or received from other components.
- Maintain a file on each isolated individual until recovery has been completed.
- Forward, but do **not** destroy, all files and the database regarding isolated personnel, their status, and/or location to the JPRA, via the JSRC, once the recovery mission is complete and the JFC no longer has a requirement to maintain the files.

FM 3-05.231

(CLASSIFICATION)			
ISOLATED PERSONNEL REPORT (ISOPREP) *(See Privacy Act Statement on reverse before completing this form.)*	1. NAME *(Last, First, Middle Initial)* DOE, JOHN		2. SSN 000-00-0000
CLASSIFIED BY: AFR 64-3 AR 525-90 NWP 19-2 DECLASSIFY ON: OADR	INSTRUCTIONS Items 1 through 15 and 20 through 23 are to be completed by Applicant. Items 16 through 19 and Item 24 are to be completed by RCC personnel. All items are to be filled in INK; however, use a pencil for items 3, 13, 14, and 20 through 24.		3. RANK/GRADE MSG/E-8
4. BRANCH OF SERVICE ARMY	5. NATIONALITY US	6. DATE OF BIRTH *(YYMMDD)* 720128	7. OBVIOUS MARKS *(Scar, Birthmark, Mole)* SEE BLOCK #24
8. BLOOD GROUP A POS	9. HEIGHT 69"	10. COLOR OF EYES BRN	11. COLOR OF HAIR BRN
12. DATE PREPARED *(YYMMDD)* 010911	13. DATE REVIEWED *(YYMMDD)* AND CURRENT ASSIGNMENT 010911	14. AUTHENTICATION NO. 8569	
		15. SIGNATURE	
16. DATE MISSING *(YYMMDD)*	17. LOSS POSITION	18. PRIORITY *(Holds vital information requiring priority rescue)* YES NO	19. SPARE

------Fold Here------

PERSONAL AUTHENTICATION SATEMENTS	
20. IN MY SPARE TIME DURING THE 1970'S, I WAS A VOLUNTEER FOREST FIRE FIGHTER FOR THE KEEP VIRGINIA GREEN CREW.	21. EVERY YEAR, I HUNTED DEER WITH MY FATHER, HARRY, ON LITTLE NORTH MOUNTAIN.
22. I FIRST MARRIED MY HIGH SCHOOL SWEETHEART, JENNY, BUT WE DIVORCED IN 1986.	23. THE HIGHLIGHT OF MY LIFE WAS THE BIRTH OF MY DAUGHTER AT FAUQUIER COUNTY HOSPITAL.
24. OBVIOUS MARKS/SCARS: • 3" SCAR ON RIGHT ABDOMEN; SCAR IN RIGHT EYEBROW. • KNOWN MEDICAL CONDITIONS OR ALLERGIES: NONE. • BOOT SIZE: 10R • CHIT# 123456789	

CONFIDENTIAL (WHEN FILLED IN)
DD 1833 [2/1(A) Electronic Version, 99 DEC] PREVIOUS EDITION IS OBSOLETE
( )
CLASSIFICATION

Figure 2-5. Example of ISOPREP

FM 3-05.231

CONFIDENTIAL (WHEN FILLED IN)
(_____)
CLASSIFICATION
EXAMPLE ONLY NOT CLASSIFIED _____ EXAMPLE ONLY NOT CLASSIFIED

AUTHORITY: 10 , Title U.S.C. Sections; 133, 3012, 5031, and 8012: EO 1397

PRINCIPLE, PURPOSES(S): It is essential to the combat search and rescue effort for the protection of search and rescue forces from enemy entrapment. The social security number is used to ensure positive identification.

ROUTINE USE(S): It is completed by each aircrew member who may be subject to action in or over hostile territory. It contains personal information that may be used to ensure positive identification. After the aircrew member has completed the form, it will be classified "Confidential."

DISCLOSURE IS VOLUNTARY: The information is necessary since it effects the entire search and rescue mission. The effect on individual of not providing information could be loss of crew status.

LEFT HAND	CODE	PRINT CODE	CODE	RIGHT HAND
1. LITTLE FINGER		Arch KK		1. LITTLE FINGER
		Tented Arch LL		
		Finger Loop MM		
2. RING		Thumb Loop NN		2. RING
		Whorl OO		
		Finger Missing PP		
		Finger Mutilated QQ		
		Question Uncertain YY		

---- Fold Here ----

PHOTOGRAPH (Front View)

2. MIDDLE				3. MIDDLE
4. INDEX	PHOTOGRAPH (Profile View)	TRP		4. INDEX
5. THUMB				5. THUMB

(_____)
CLASSIFICATION

Figure 2-5. Example of ISOPREP (Continued)

2-9

UARCC

2-14. The branch of the SOC J-3 normally responsible for compartmented operations forms the nucleus of the UARCC. The mission of the UARCC is to integrate and coordinate all theater NAR capabilities in support of the JFC's PR operations. The UARCC is a compartmented SOF facility staffed on a continuous basis by supervisory personnel and tactical planners. UARCC personnel coordinate, synchronize, and deconflict NAR operations within the operational area assigned to the JFC. The UARCC interfaces and coordinates with the JOC, the JSRC, and when directed, with each theater component RCC. (Appendix D discusses the manning and operations of the UARCC.) The UARCC fulfills similar responsibilities and functions as each theater component RCC, but tailors those roles and functions from the perspective of NAR support for the theater. Once established, the UARCC conducts the following critical tasks:

- *Advises.* The UARCC advises the commander on the development and employment of NAR capabilities in support of the theater PR plan.
- *Communicates.* The UARCC provides the connectivity between the NAR forces and the theater PR architecture to provide time-critical information. The UARCC is the conduit through which launch-and-execute authority is passed.
- *Coordinates.* The UARCC coordinates all representative UAR and NAR capabilities. It coordinates support for the comprehensive NAR ground tactical plan. When directed by the JSRC, the UARCC coordinates with theater component RCCs for conventional recovery support.
- *Integrates.* The UARCC integrates NAR capability into the theater PR plan.
- *Deconflicts.* The UARCC deconflicts NAR operations internally with all NAR forces supporting a single recovery operation. It also deconflicts a NAR operation externally with other theater operations. The UARCC conducts timely exchange of operational and support information to aid mission execution, avoid disruption of ongoing operations, and avoid fratricide.
- *Synchronizes.* The UARCC synchronizes the ground tactical plans between NAR tactical elements. Synchronization also occurs between NAR tactical elements and conventional recovery support, conventional military operations, SO, and interagency activities.

PR CELL ESTABLISHMENT

2-15. Assuming sufficient manpower, space, and secure communications systems are available, the SF operating base (SFOB, forward operational base [FOB], or advanced operational base [AOB]) commander may elect to establish the functional equivalent of an RCC. This functional equivalent will be dedicated exclusively to the planning and operations of SF PR.

SF COORDINATION FOR PR

2-16. PR, as it applies to SF, is operational in nature. Therefore, SF PR coordination must be resident in the OPCEN. The OPCEN should ensure clear, expeditious lines of communication to the JFSOCC's RCC (SOC RCC) to manage CSAR with respect to SF and their direct support to JCSAR. Additionally, OPCEN personnel with access to detailed mission-specific information related to UAR and NAR must possess the requisite security clearances (generally a Top Secret [TS] and Sensitive Compartmented Information [SCI]) and should be formally trained to employ, supervise, and manage advanced SO resources. The OPCEN, usually via an operational control element (OCE), must ensure clear, expeditious, and compartmented lines of communication to the UARCC.

SF PR COORDINATORS

2-17. SF must view PR from the dual perspective of potential consumer and potential force provider. SF PR coordinators are OPCEN staff officers and noncommissioned officers (NCOs) tasked to manage SF PR matters. They are responsible to plan, coordinate, and track all SF PR operations. SF PR coordinators should be formally trained; for example, the JPRA Personnel Recovery Plans and Operations Course, PR301. SF PR coordinators should also have a thorough understanding of theater PR architecture, roles and missions, and coordination process. The OPCEN logistics planner, OPCEN signal planner, and the OPCEN operations planners must have periodic meetings to ensure resources are available to meet PR mission requirements.

2-18. PR planning such as intelligence preparation of the battlespace (IPB) and operational preparation of the battlespace (OPB) is a continuous process. SF PR coordinators must be familiar with all aspects of PR, to include SERE, CSAR, JCSAR, and NAR. PR planning is inherent to force protection, and recovery planners should address PR in all missions. All planning should be based on the five tasks of PR (report, locate, support, recover, and return or repatriate). SF PR coordinators must assume that every mission may result in an evasion situation and should prepare the SFODA for unassisted evasion. Plans for both assisted and unassisted evasion must be developed IAW theater PR guidance to ensure a coordinated recovery of an SFODA by PR forces. Within the SFOB or FOB, the SF PR coordinators should be task-organized and prepared to fulfill their roles and responsibilities in support of PR planning and coordination for SF operations. PR planners must consider the availability of resources and the capabilities and limitations of the resources and task-organize accordingly.

2-19. SF PR coordinators should help SFODAs develop and refine their evasion SOPs. SF PR coordinators should provide guidance, quality control, and coordination as SFODAs develop their EPAs, including their "stand alone" EPA overlays. This coordination ensures thorough PR planning as the SFODA builds its EPA. SF PR coordinators are also responsible for the coordination, planning, and oversight of all SERE, CSAR, and UAR training, to include schools and exercises.

C2 FOR SF AS THE FORCE PROVIDER

2-20. As part of the integrated PR system of the theater, the SFOB commander has the command responsibility for CSAR of his forces consistent with his organic capabilities and assigned functions and IAW requirements of the supported combatant commander. The SF PR coordinator(s) within the OPCEN should establish secure and reliable connectivity with the SOC RCC.

The SF commander should have his assigned forces, tasked to support CSAR, prepared and in an appropriate readiness posture to respond to CSAR mission taskings as assigned by the JFSOCC and communicated via his RCC.

2-21. Since SF operations are normally conducted in the deep battlespace, the SFOB commander's ability to conduct CSAR in support of his own operations is generally very limited. If the SFOB commander is tasked by the JFSOCC to provide forces to support CSAR, the resulting OPCON-TACON relationship will be delineated with ARSOF or AFSOC commanders. Since CSAR is a dynamic, time-sensitive event, SFODAs supporting a CSAR mission can expect an extremely abbreviated joint or combined mission planning sequence, which is atypical of standard SF operations.

2-22. To control UAR operations, the SFOB commander normally retains OPCON of his deployed unconventional assisted recovery teams (UARTs) and unconventional assisted recovery mechanisms (UARMs), unless the JFSOCC specifically retains that authority. The SF PR coordinator, normally a part of the OCE, is the conduit through which tactical planning for UAR and SAR message traffic is transmitted between tactical planners in the UARCC and deployed SFODAs. The SF PR coordinator requires secure communications with the UARCC and with the deployed UARTs and UARMs. Moreover, he must maintain detailed visibility of the capability, limitations, and infrastructure development of each SF UART and UARM, while ensuring the compartmentation of each.

PREMISSION PLANNING

2-23. The PR coordinators should establish contact with the SOC RCC and solicit theater PR guidance. PR coordinators should acquire the following:

- Theater SERE and high-risk-of-capture guidance.
- Isolated personnel guidance.
- SERE contingency guides.
- Joint Personnel Recovery Support Products (JPRSPs).
- ATO.
- Theater CSAR CONOPS.
- CSAR SPINS.

2-24. To facilitate the development of the EPA for the SFODA, PR coordinators should extract pertinent information such as the following:

- Letter, number, color, and word of the day.
- Duress word.
- Search and rescue dot (SARDOT).
- Search and rescue numerical encryption grid (SARNEG).
- Ground-to-air signals (GASs).
- Primary and alternate SAR radio frequencies and procedures.
- Established spider points and routes.

2-25. The SF PR coordinators, coordinating with the SOC RCC, should prepare and disseminate an operations graphic depicting range and time of theater CSAR capabilities. The PR coordinators should disseminate the determination by the theater of the SFODA legal status, any status-of-forces agreements (SOFAs), and rules of engagement (ROE) relative to PR. PR coordinators should inform the SFODA of any PR-related PSYOP preparation of the battlespace.

2-26. In preparation to coordinate and conduct PR operations, SF PR coordinators within the OPCEN should prepare their workspace to manage SF PR operations. SF PR coordinators should periodically confirm (with the SOC RCC) the currency of on-hand ATO, CSAR CONOPS, and CSAR SPINS and the validity of data that may become compromised. SF PR coordinators use the following tools to aid the management of PR missions:

- A PR asset board (Figure 2-6).
- A PR incident/mission board (Figure 2-7, page 2-14).
- An operations map with threat overlays and a CSAR operations graphic overlay that includes the established SARDOT(s), spider points and routes, and PR mission folders and worksheets.
- A posting of the letter, number, color, and word of the day; duress word; and SARNEG.
- GASs.

Higher, Adjacent, or Other Elements	POC or Address and COMM Plan	Equipment Assets	Response Time	Capability Limitation	Current Status	Notification Time
JSRC						
U.S. EMBASSY						
HOST NATION						
UARCC						
SOC RCC						
COMPONENT RCCs						
NATIONAL AGENCIES						

Figure 2-6. PR Asset Board

2-27. In anticipation of need, PR coordinators should be prepared to provide evasion aids to support deploying SFODAs. Evasion aids include evasion charts (EVCs), blood chits, pointee-talkees, evasion kits, survival radios, and other signaling devices. Signaling devices should be redundant, active and passive, day and night, visible, IR, and multispectral. SF PR coordinators should acquire blood chits and other evasion aids through the SFOB staff and SOC RCC from the theater PR OPR or JSRC.

Incident/Mission Board										
Inc No/ DTG	Msn No/ DTG	Type of Inc	Name or Call Sign	Unit	No of Pax	FRAG Order Msn No	Position/ LKP	Threat	Time of Last Contact	Status/ Remarks

SARDOT: WORD OF THE DAY:

SARNEG: NUMBER OF THE DAY:

LETTER OF THE DAY:

COLOR OF THE DAY:

Figure 2-7. PR Incident/Mission Board

2-28. Many of these items may also be required for tactical operations. The OPCEN director must know the status of these assets and be able to allocate them according to the commander's priorities. PR coordinators should track the availability and usage of these items on a chart in the OPCEN. The PR coordinators should also be aware of organic, SOC RCC, and JSRC-coordinated recovery assets potentially available to support recovery of deployed SFODAs.

PR MISSION PLANNING

2-29. SO conducted by SF are often conducted in uncertain or hostile areas at great distances from operational bases and friendly support. Regardless of mission profile (SR, DA, UW, and so on), SF missions require detailed planning, to include insertion, resupply, fire and maneuver support, extraction, and evasion. Therefore, all SFODAs must plan and prepare for unassisted evasion.

2-30. The SFOB planning staff will assemble to begin mission planning. The SF PR coordinators in the OPCEN will assemble with the planning cell and participate throughout the planning phase of the operation. The SF PR coordinator is responsible during mission planning to help the SFODA develop and coordinate its EPA.

MISSION ANALYSIS

2-31. The first step of the military decision-making process (MDMP) is mission analysis. During mission analysis, the PR coordinator should ensure he has a thorough understanding of the mission and the commander's intent. If it has not already been accomplished, the PR coordinator should conduct his mission analysis with a PR perspective and IAW PR guidance that may have been provided separately by the SOC RCC and JSRC. The PR coordinator should extract specified and implied tasks related to PR from the tasking order (TASKORD) and develop a timeline for PR planning support that is consistent with the timeline of the SFOB planning cell.

PR INTELLIGENCE ANALYST

2-32. While the staff is conducting mission analysis, the SFOB intelligence section works on its intelligence estimate. The intelligence analyst supporting PR should participate in this process. Intelligence specialists within the OPCEN, tasked to support SF PR, play a key role in planning and management of SF PR operations. SF intelligence planners should provide general information such as a country study containing the following:

- The behavior and attitudes of the local populace and resistance organizations.
- Indigenous culture and taboos.
- Special PSYOP studies.
- Intelligence summaries.
- Rewards or bounties.
- SOFAs.

2-33. The intelligence analyst should also provide information specific to PR, to include the following:

- Information and overlays of enemy ground and air threats.
- Enemy counter-CSAR capabilities.
- Active and passive security.
- Population control measures (curfews, rationed items, security hazards, checkpoint procedures, and so on).
- Potential support from the local population, to include barter items.
- Border crossing procedures.
- Potential enemy captivity and interrogation procedures.

2-34. PR intelligence analysts are responsible for the following:

- IPB specific to PR, to include assessment of the threat environment. Analysts should identify, as a minimum, the following threats in the area of operations (AO):
  - Threat security forces (active security measures, passive security measures, security force composition, border crossing procedures, and documentation requirements).
  - Special capabilities (counterintelligence, electronic measures, ground surveillance radar, SOF, night vision devices, aerial and satellite surveillance capabilities, and suspected interrogation techniques).
  - Effects of noncombatants on PR.
- Development and processing of requests for information (RFIs) specific to PR for the SOC RCC and the J-2.

- Development of an intelligence overlay of pertinent enemy order of battle (OB).
- Development of modified combined obstacle overlays (MCOO) and doctrinal, situational, and event templates.
- Participation in PR war-gaming and support during development of the decision support template (DST).
- Preparation for the tactical intelligence debriefing of SFODA members and extraction of any information related to PR.

2-35. The PR intelligence analyst must create an overlay that graphically depicts all evasion data. Sources of this information include, but are not limited to, the following:

- JPRSPs.
- A selected area for evasion (SAFE) area intelligence description (SAID).
- Other areas identified by the theater to support PR; for example, designated areas for recovery (DARs) or selected contact areas.

This overlay is combined with an MCOO and doctrinal, situational, and event templates. Graphic representation of existing PR data, including any predesignated evasion corridors, coupled with the enemy ground and air threat and friendly force locations, allows the PR coordinator to more readily visualize possible PR COAs. The PR intelligence analyst must participate in the IPB process to integrate intelligence into the EPA development process.

2-36. The intelligence analyst must determine the impact on and expected behavior of civilians in the AO. The intelligence analyst must also view from a PR perspective projected friendly and enemy COAs in terms of time and space. By doing so, the intelligence analyst can determine any impact on anticipated PR operations, to include the potential of unassisted evasion of an SFODA.

PLANNING STAFF COORDINATION

2-37. When the planning cell completes its mission analysis, it will brief the SFOB commander. The briefing should begin with the S-2 intelligence estimate. This estimate will lay the foundation for follow-on COA development. The S-2 should also present his recommended PIR, which may include PR requirements identified by the PR coordinators. The S-3 will next present the staff mission analysis. The staff mission analysis should include the following:

- Specified and implied tasks.
- Mission-essential tasks.
- Constraints.
- Restrictions.
- Timeline.
- Risk assessment.
- Feasibility analysis.
- Restated mission.

2-38. The staff may also update the commander on availability of resources to support the mission and the recovery of the SFODA.

2-39. Following this initial briefing, the SFOB commander will articulate his own independent analysis of the mission and provide planning guidance. His planning guidance should include abort and evasion criteria. His planning guidance may also include levels of acceptable risk and recommended COAs for the staff to analyze.

2-40. Following the first briefing, the staff will begin COA development. The PR coordinator should help with COA development by identifying potential PR support assets within the theater for each COA and by focusing his efforts on synchronizing PR planning with comprehensive mission planning. The PR coordinator should coordinate with the PSYOP officer in the OPCEN to assess support to PR. PSYOP themes that support PR and complement the JFC's PSYOP campaign plan may prove useful by influencing the local populace to either assist or refrain from interfering with potential evaders. (Chapter 5 includes more information on this subject.)

2-41. After the staff has developed multiple COAs, it must analyze each one by using the evaluation criteria specified by the SFOB commander. The battlefield operating systems (BOS) are excellent tools for evaluating each COA. The SF PR coordinator should also analyze each PR COA from a BOS perspective. The SF PR coordinator must provide the PR analysis of each COA to the OPCEN director to define PR supportability and aid the COA analysis process.

2-42. Part of the COA analysis will be war-gaming. The proposed PR plan should be war-gamed using the same methodology as the proposed COAs. The most popular methods of war-gaming are belt, box, and avenue in depth. The belt method provides an excellent tool for war-gaming a PR plan. It allows the war-gamer to focus on specific events and actions (action, reaction, and counteraction). The SF PR coordinator acts as the friendly forces, and the PR intelligence analyst or S-2 acts as the enemy. During this step of the MDMP, the PR DST is developed. The PR DST is an overlay that graphically depicts the war-gaming process. This overlay should include all enemy, friendly, noncombatant, and PR data and MCOO graphics.

2-43. Also during war-gaming, the SF PR coordinator should develop an evasion support timeline (EST). The EST is determined from the PR DST. The PR coordinator must determine the decision points requiring decisions from the SFOB or SFODA commander. Example decision points include the following:

- After no radio contact, when should an SFODA or individual be considered as evading?
- When should an emergency resupply bundle be transported to the airfield?
- When should the SFODA begin movement to the resupply drop zone (DZ)?
- When should a restricted operations zone (ROZ) be coordinated, established, and deactivated?

2-44. It is doubtful that the entire evasion corridor will be a ROZ. The SF PR coordinator must be able to visualize the EST and be able to coordinate the activation and deactivation of applicable ROZs. The SF PR coordinator must also advise the staff on resupply options that would facilitate the continuation of the SFODA mission and/or affect its recovery. Some events that should be included on the EST are as follows:

- Earliest anticipated launch time (EALT).
- Target window.
- Critical decision points for the SFOB or SFODA commander.
- Not later than (NLT) times:
    - NLT for SFODA to establish initial communications.
    - NLT time to notify SOC RCC of a potential CSAR mission.
    - NLT time to drop emergency resupply.
    - NLT time to deliver resupply bundle to airfield.
    - NLT time to notify resupply or recovery aircraft.
    - NLT time to submit SARIR after liaison officer's (LNO's) contact with element.
    - NLT times to activate or deactivate ROZs.
    - NLT time to notify SOC RCC of a potential CSAR mission and transmit required EPA and ISOPREP information.
- Earliest anticipated exfiltration (exfil) times.
- EST matrix (Figure 2-8). This matrix allows the coordinator to visualize time and space in relation to activities of friendly forces and enemy forces.
- Sequence of events timeline for recovering an SFODA with organic assets. Figure 2-9, page 2-19 shows an EST.

Infil 01020Z	SR Target Window 02001Z-04001Z			Exfil 042200Z
	Day 2	Day 3	Day 4	
Immediate contact; SFODA exfilled by infil helo	SFODA contact window 01			
		SFODA contact window 020800Z		NLT contact with SFODA 030400Z (infil + 50 hours). Submit SARIR to SOC RCC	Drop emergency resupply bundle 040200Z
			SFODA contact window decision point for SFODA commander (whether to go to resupply DZ or continue with mission) 022200Z		SUPCEN transports bundle to airfield 040100Z

Figure 2-8. Sample EST Matrix

FM 3-05.231

N hour	FOB determines SFODA is conducting E&R. FOB submits SARIR to SOC RCC. FOB determines if it can recover the SFODA with its own assets. SUPCEN assembles and rigs resupply bundle. SF PR coordinator transmits EPA to SOC RCC. SF PR coordinator reviews EPA and E&R overlay with OPCEN.
N + 1	OPCEN director briefs FOB commander on situation and recommended objective area (OA), organic recovery team notified, mission preparation begins.
N + 2	Finalize organic recovery plan, and brief SOA pilots.
N + 3	SUPCEN transports resupply bundle to airfield.
N + 4	SOA ensures route deconfliction through JSOAC, SOC RCC, and JSRC. Organic CSARTF briefs FOB commander. All RCCs, SOC RCC, and JSRC notified of impending CSAR mission. Organic CSARTF conducts final rehearsals and inspections.
N + 5	Organic CSARTF launches, if METT-TC permits and only after launch authority has been granted by the appropriate HQ; that is, JTF, JFSOCC, JFACC, or JSRC.
NOTE: N hour occurs 24 hours after infiltration or 24 hours after a missed communications window.	

Figure 2-9. Sample EST

2-45. Following the COA analysis, the SFOB staff will compare COAs to determine a recommended COA. The PR coordinator should provide input on PR supportability of each COA. The planning staff and OPCEN director consider the SF PR coordinator's input when comparing various COAs. Normally, the planning staff uses a decision matrix to facilitate this process.

2-46. The planning staff will present a decision brief on the recommended COA, usually including the decision matrix, to the SFOB commander. The SFOB commander will approve, approve pending changes to, or disapprove the COA. The planning staff should also present the PR plan to the SFOB commander for his final approval. Additionally, the PR coordinator should reiterate the abort and evasion criteria and ensure its inclusion in the OPORD. Although the commander's decision is the final step of the SFOB MDMP, it is not the end of this particular phase of planning.

2-47. The SF PR coordinator must ensure the required PR data are included in the appropriate sections of the OPORD. Additionally, the SF PR coordinator must build the shell of the EPA with information derived from the theater-specific PR references; for example, theater SERE and high-risk-of-capture guidance, isolated personnel guidance, applicable SERE contingency guides, area-specific JPRSPs, and theater-specific ATO, CSAR CONOPS, and CSAR SPINS. By completing portions of the EPA from PR references and intelligence data, an isolated SFODA can use its isolation time for more important activities, such as rehearsals.

2-48. Upon completion of the OPORD (to include the EPA annex), the OPCEN will provide a designated SFODA with its isolation packet. This packet should include the isolation schedule, OPORD, copies of the SFOB MDMP, copies of theater-specific PR references, and copies of the PR coordinator's MDMP. This packet should be provided to the SFODA as soon as possible to allow the SFODA to digest the available information and prompt appropriate questions from the SFODA during the staff mission brief.

2-49. The SFOB staff will present a staff mission briefing to the designated SFODA. This briefing will include all staff sections of the SFOB. During the briefing is also the first opportunity for the SFODA to address the SFOB commander and his entire staff. Following this briefing, the SFOB commander will articulate his intent, specific planning guidance, acceptable risk, and specific rehearsal requirements. The SFOB commander may include portions of the PR plan in the rehearsal requirements for the SFODA.

2-50. At this point, the SF PR coordinator and the LNO of the isolated SFODA become the focal point for the SFODA PR planning. The LNO, who participated in the SFOB MDMP, provides continuity between the OPCEN and the isolated SFODA. The SFODA conducts its own MDMP similar to that of the SFOB staff. The SFODA completes its EPA, tailored to the SFODA and its assigned mission. If the SFODA intends to infiltrate by air, the SFODA, facilitated by the LNO and the PR coordinator, should coordinate PR contingency plans with the infiltration air mission commander and crew. The SFODA also ensures ISOPREP cards are complete and current for each deploying team member. The EPA and ISOPREP cards of the SFODA should be secured together to facilitate coordination for recovery of the SFODA should it become necessary. The LNO should be intimately familiar with the plan of the SFODA.

2-51. The briefback is the key piece. The commander must be thoroughly briefed on the EPA. The commander must "know" that the EPA is "coordinated" and "supportable." If only one portion of the SFODA mission can be briefed, it should be the EPA. The SFODA must be sure to "inform" the commander that its EPA is "our plan." It is not "set in stone." EPA phase lines (PLs) may or may not be met by the projected timeline. The SF PR coordinator must be present at the briefback.

2-52. Following the SFODA infiltration, the LNO becomes part of the OPCEN and tracks the progress of the SFODA. This places the LNO in the best position to react to potential evasion situations and interface with the PR coordinator.

MISSION EMPLOYMENT

2-53. Given the operational environment in which SF units conduct combat operations, any deployed SFODA may be forced to execute its EPA. SF PR coordinators must maintain visibility over all deployed SFODAs and anticipate evolving PR events.

2-54. A PR event is initiated with a SARIR (Figure 2-10, page 2-21). SARIRs are communicated in near-real-time between the JSRC and theater component RCCs. In turn, the SOC RCC should transmit the SARIR to the SF PR coordinators of all SOF components for situational awareness.

SARIR
LINE 1: CALL SIGN
LINE 2: TYPE (aircraft, ship, or other)
LINE 3: COLOR
LINE 4: ID (aircraft tail number, hull number, and so on)
LINE 5: LOCATION (geographic coordinates, UTM, and so on)
LINE 6: QUALIFIER (actual or estimated)
LINE 7: TIME (DTG + zone of incident)
LINE 8: CAUSE
LINE 9: PERSONNEL (actual or estimated)
LINE 10: STATUS (count and condition)

Figure 2-10. Sample SARIR

2-55. Recovery of isolated personnel is normally a very dynamic event that requires decisive action. Upon receipt of a SARIR, SF PR coordinators should immediately conduct a capability and feasibility analysis to determine how the SFOB could support the recovery mission. If the SF PR coordinator determines that the SFOB can support a particular recovery, he should communicate that capability and feasibility to the SOC RCC to be incorporated into the integrated and synchronized recovery plan. That capability may include the opportune use of deployed SFODAs, organic CSAR in support of SF operations, or UAR. As additional information becomes available, the JSRC will disseminate SARSITs (Figure 2-11) via the theater PR architecture. Therefore, the SFOB OPCEN must maintain connectivity with the SOC RCC.

SARSIT
LINE 1: MISSION NUMBER
LINE 2: STATUS (of recovery mission)
LINE 3: CALL SIGN (of evader)
LINE 4: TYPE (aircraft, ship, or other)
LINE 5: LOCATION (geographic coordinates, UTM, and so on)
LINE 6: PERSONNEL
LINE 7: PERSONNEL STATUS
LINE 8: NARRATIVE
LINE 9: TIME
LINE 10: AUTHENTICATION

Figure 2-11. Sample SARSIT

2-56. If the PR capability provided by SF is determined by the SOC RCC to be the best force option, the OPCEN will receive a search and rescue request (SARREQ) via the SOC RCC directing mission execution. The SARREQ message requests forces to participate in a CSAR mission. This message is normally sent from the JSRC to component RCCs and any designated functional commanders. It records arrangements made to employ resources from two or more components to prosecute a CSAR mission.

2-57. Once an SFODA infiltrates its assigned AO, its LNO becomes part of the OPCEN and tracks the current operations of the SFODA. The information obtained from the SFODA through the LNO is critical to the SF PR coordinator. The PR coordinator must carefully match the progress of the SFODA with his EST. He should continuously war-game potential problems and proactively support the potential recovery of the deployed SFODA, if required.

2-58. The SF PR coordinator must continuously update his status boards. He should also maintain good communications with the SOC RCC and be aware of any PR activities in the SFOB commander's area of interest.

SF AS THE CONSUMER

2-59. The SF PR coordinator should notify the SFOB commander and the OPCEN director before each evasion decision point is met. The PR coordinator should request guidance on whether to proceed with the recovery activities identified on his EST or modify them according to the friendly or enemy situation.

2-60. If a deployed SFODA reports contact with the enemy, the SF PR coordinator should conduct planning as if the SFODA will eventually initiate its EPA. The SF PR coordinator should develop timeline templates that facilitate reporting and requesting support to aid recovery of the SFODA. Doing so will facilitate the quickest recovery of evaders, if required.

2-61. Once an SFODA initiates its EPA, the PR coordinator must react decisively and keep the entire chain of command informed. After notifying the OPCEN director and SFOB commander, the PR coordinator must notify the SOC RCC via a SARIR. He also informs the SOC RCC of the current capability of the SFOB to recover the SFODA. If the SFOB is unable to recover the evading SFODA using organic CSAR forces, the SOC RCC will then determine if it possesses the necessary resources to recover the isolated personnel. If it does not, the SOC RCC will request assistance from the JSRC. Throughout the PR operation, the SF PR coordinator updates the SOC RCC with SARSITs for dissemination throughout the PR architecture.

2-62. During the PR event, the SF PR coordinator tracks the progress of the SFODA through the SFODA LNO and via information provided by the SOC RCC and JSRC. The SF PR coordinator must continuously war-game the actions of the evading SFODA and anticipate their requirements. The PR coordinator must anticipate the requirements for additional assistance and request augmentation, if required. Throughout the PR event, the PR coordinator must not neglect the PR requirements of other deployed SFODAs operating in their assigned AOs.

SF AS THE FORCE PROVIDER

2-63. SF participation in a PR event could be conducted as either SF CSAR, SF in support of a CSARTF, or a UAR event. CSAR or JCSAR is a dynamic event that must be coordinated throughout the theater PR architecture. CSAR or JCSAR should be executed as a deliberately planned "on order" mission with a standing force in a requisite readiness posture measured in minutes. OPCON relationships must be clearly defined. Once the SFOB OPCEN receives the SARREQ, the SFOB commander directs the execution of the specified PR event as coordinated with the JSRC via the SOC RCC. SFOB PR coordinators must remember that EPAs and ISOPREPs must also be prepared for all CSAR forces.

2-64. The SF PR coordinator should know the location of all deployed SFODAs. The coincidental location of an SFODA near emerging PR events may result in the SFODA supporting the PR event. If a deployed SFODA could potentially support a PR event, the SF PR coordinator should notify the SOC RCC, highlighting that the use of a deployed SFODA could severely impact the ability of that SFODA to continue its original mission. The JFSOCC normally has command authority to divert an SFODA to opportunistically support an emerging PR event.

2-65. The SF PR coordinator should know the location and progress of PR events involving JCSAR supported by SF soldiers. He can then keep the SFOB commander informed of the current status of his soldiers. The SF PR coordinator must also know the location and progress of deployed SF UARTs and UARMs. He is the conduit through which SARIRs, SARSITs, and SARREQs are passed between the UARCC and the deployed UARTs or UARMs. He is also the conduit through which UARTs and UARMs and tactical planners within the UARCC coordinate, synchronize, and deconflict UAR missions. The SF UARTs and UARMs also transmit support requirements related to the UAR mission to the UARCC.

POSTMISSION

2-66. Before the return of SFODAs involved in a PR event, the SF PR coordinator must organize and coordinate SFOB staff support beginning at the recovery airfield. He must provide oversight of staff support, to include tactical intelligence debriefings, operational debriefings, and medical and logistical support. For returning SF evaders, SERE debriefings conducted by theater SERE specialists may be required as well as NAR debriefings conducted by JPRA, when appropriate. The SF PR coordinator must ensure preparations are IAW theater repatriation plans.

2-67. Repatriation is the final task of PR. The intent is to return the evader back to duty as soon as possible. Repatriation is a four-phase operation controlled by the chain of command. DODI 2310.4, *Repatriation of Prisoners of War (POW), Hostages, Peacetime Government Detainees, and Other Missing or Isolated Personnel* discusses the four phases. Phase 1 (Initial) normally occurs near the recovery airfield. Phase 1 generally addresses the immediate actions of medical screening and initial tactical intelligence and operational debriefings.

2-68. Medical concerns will generally take precedence over other aspects of repatriation. If soldiers are wounded or suffering from exposure or dehydration, an ambulance and medical assistance should be positioned at the airfield. Many aspects of Phase 1 of repatriation can occur within the secured confines of a medical facility. PR coordinators should coordinate with the SOC RCC to support Phase 1 and Phase 2 (Transition) to provide support to recovered SF soldiers.

2-69. Phase 2 of repatriation occurs when the commander, based on the input from key staff, determines the returning soldier requires additional support. Phase 2 normally occurs at a designated theater site IAW the repatriation plan of the theater. The geographic combatant command, supported by the service components, will coordinate IAW DODI 2310.4. The returning evader can expect more tactical intelligence, operational, and SERE debriefings. If the isolated personnel are recovered via UAR or NAR, debriefings specific to NAR support are reserved for representatives of the NAR division of the JPRA.

2-70. Phase 3 (Detailed Processing) of repatriation occurs when the commander determines that the returning soldier needs more support than is available within the theater and is normally conducted at a designated continental United States (CONUS) facility. The theater command passes responsibility for Phase 3 to the parent Service IAW DODI 2310.4. The parent Service is also responsible for Phase 4 (Long-Term Rehabilitation of Returnee).

NOTE: Throughout the evasion, recovery, and repatriation process, the SF PR coordinator should plan and coordinate necessary support for the evader's family.

2-71. Once repatriated personnel are returned to the SFOB, SF PR coordinators should prepare for and participate in the SF tactical-level debriefing process and incorporate lessons learned into the SFOB PR capabilities. Generally, these debriefs will involve tactical intelligence and operational debriefing. The SFOB should apply any information obtained from debriefings and lessons learned to improve SF training, planning, equipping, and conducting of operations.

Chapter 3
Personnel Recovery for Special Forces

SF missions are usually dangerous, thereby increasing the chances SF personnel may become isolated. As such, SF personnel must complete SERE training, complete an ISOPREP (DD Form 1833), and develop an EPA. SF personnel must be familiar with the tools for evasion; for example, pointee-talkee, EVC, blood chit, and so on. Because assistance is not always immediately available, SF personnel must be prepared to evade and arrive at a location where they can be rescued.

SF personnel have unique skills that make them effective PR assets. SF units are responsible for CSAR in support of their own personnel who become isolated. The SFODA is the basic element of the SF group. In the PR arena, the SFODA can be used and tailored to accomplish the following PR missions: CSAR, JCSAR, UAR, and NAR.

SECTION I – UNASSISTED EVASION: SPECIAL FORCES AS THE EVADER

3-1. SO performed by SF are often conducted at great distances from operational bases and friendly support in hostile, denied, or politically sensitive areas. During the conduct of these SO, SF personnel are often placed at great risk and are subject to becoming isolated in hostile areas. Personnel should be prepared to survive and evade over long distances for an extended period of time (more than 48 hours) to a location where recovery may be effected or make their way to friendly or neutral territory without assistance. Isolated personnel should remember that evading is always better than capture. SF personnel must consider unassisted evasion during operational planning as a contingency in the absence of assisted recovery options.

UNASSISTED EVASION

3-2. The following vignette describes Chris Ryan's (Bravo Two Zero) unassisted evasion. Figure 3-1, page 3-2, shows a map of the general area in which the evasion occurred and the path Ryan took to evade and ultimately escape.

On the night of 22 January 1991, at a remote airfield in Saudi Arabia, under the cover of darkness and in conditions of the utmost secrecy, eight members of the British Special Air Service (SAS) Regiment, call sign Bravo Two Zero, boarded a U.S. Chinook helicopter. Their mission was to infiltrate deep behind enemy lines and relay intelligence on troop movements, destroy Iraqi fiber-optic communications systems, and to seek and destroy mobile surface-to-surface missile system launchers before Israel was provoked into entering the war. Each man was laden with over 150 pounds of equipment. They patrolled 20 kilometers (km) (12.4 miles) across flat desert to reach their objectives and find a hiding place before first light.

FM 3-05.231

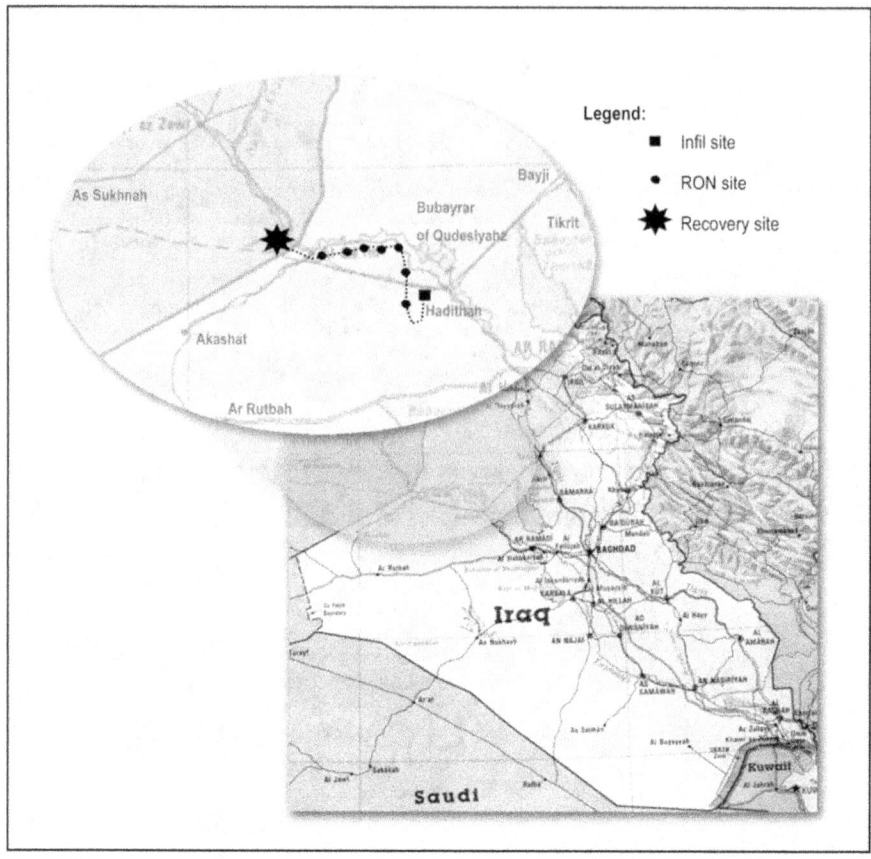

Figure 3-1. Bravo Two Zero—Chris Ryan's Successful Unassisted Evasion Route

The mission went sour from the outset. Their radios did not work, and the weather turned cold enough to freeze diesel fuel. Worst of all, they had been spotted by a boy who was herding his goats. They decided to let the boy live and make a run for it, but were tracked down and soon surrounded by Iraqi forces. The Iraqis attacked with armor, and after a fierce firefight, the patrol was forced to evade on foot toward the Syrian border, 120 km (74.6 miles) to the northwest. To assist them in their evasion route planning, they were forced to rely on information contained on escape maps that were printed in 1928 and then updated for WW II.

During the first night, the patrol accidentally broke into two groups of five and three. Chris Ryan found himself left with two companions. Nothing had prepared them for the vicious cold of the desert winter, and they began to suffer from hypothermia. During the night, one of the men disappeared in a blinding blizzard. The next day a goat herder walked on top of the two survivors. Chris's remaining partner, thinking the Arab friendly, went off with him in search of food, and never returned.

Left on his own, Chris Ryan set out to continue his escape toward the border. Soon he began to have trouble because he had no food or water left. On Day 7, he came to a river and drank from it, but he did not know it was polluted with chemical waste. He fainted a couple of times during his march to the border, and visions of his daughter walking ahead of him were one of the things that kept him going. When he had almost reached the border, he suddenly realized that he had just entered a military camp. Two men approached him, and he had to kill them with his combat knife.

In the end, he managed to reach the border and make it to the British embassy in Damascus, Syria. Chris reached the Syrian border after 8 days. He had walked more than 270 km (167.8 miles) in a standard issue military desert suit, survived temperatures close to –20 degrees Celsius (-4 degrees Fahrenheit) on average, and had eaten nothing but a small pack of biscuits. He walked during the night and hid during the day. The water had infected his lips. It took him 2 weeks to be able to walk again and 6 weeks to regain sensation in his fingers and toes. Chris was awarded the Military Medal for his perseverance during the evasion.

Sources: Andy McNabb, *Bravo Two Zero*, 1993.
Chris Ryan, *The One That Got Away*, 1995.

TRAINING

3-3. SF isolated personnel have the responsibility to assist in their own recovery to the maximum extent possible. Successful unassisted evasion is dependent on effective predeployment and premission preparation and training.

POLICY

3-4. USSOCOM policy requires all operational forces to complete SERE Level C training with the priority going to national mission and major theater of war (MTW) apportioned forces followed by all remaining personnel, as required. Furthermore, geographic combatant commanders have defined minimum training requirements for entry into their theater of operations.

BACKGROUND

3-5. Department of Defense Directive (DODD) 1300.7, *Training and Education to Support the Code of Conduct (CoC)*, 8 December 2000, mandates that "DOD personnel who plan, schedule, commit, or control the use of the Armed Forces shall fully understand the CoC and ensure that personnel are trained and educated to support it." Code-of-Conduct and SERE training provide soldiers the necessary skills to survive and evade in a hostile environment, to conduct themselves appropriately during captivity, and to return with honor. The Code of Conduct is applicable across the spectrum of conflict, to include wartime captivity, peacetime governmental detention, and during hostage situations.

3-6. DODI 1300.21, *Code of Conduct (CoC) Training and Education*, 8 January 2001, establishes three levels of training and provides guidance to support each level. The levels and applicability of SERE and Code-of-Conduct training are:

- *Level A.* This level applies to all members of the armed forces. This training focuses on familiarity with the Code of Conduct and the individual's legal and moral responsibility to comrades, the Service, and the nation. This training is given to all personnel upon entry into military service.
- *Level B.* This level applies to personnel whose military role entails moderate risk of capture and exploitation. Level B is primarily academic instruction on SERE and Code-of-Conduct basics. Practical application and/or exercise of survival, evasion, and recovery instruction is encouraged but not required. Practical application and/or resistance training during simulated captivity are prohibited. Training may be conducted in-unit or at a formal Service training location.
- *Level C.* This level applies to personnel at relatively high risk of capture and exploitation. Academic, field, laboratory, practical application, and/or exercise of all aspects of SERE and the Code of Conduct are part of the course curriculum performed under the supervision of Service-qualified instructors. Training is conducted at four formal sites around the United States.
- *Other training.* Other training programs provide specialized Code-of-Conduct and/or SERE training according to the unique requirements of the units or personnel receiving the training. Selected SF personnel are trained in the following programs:
    - SV82A, Joint Service Training Program.
    - SV83A, Special Survival Training.
    - SV91A, DOD High-Risk Survival Training.
    - SV93A, DOD Hostage Survival Training.

3-7. The intent of Levels A, B, and C and other training is to provide subject-matter guidance for use in ascending levels of understanding and to direct the military Services to increase each Service member's depth of knowledge, depending on his needs. Training at Levels B and C should include detailed information on coping skills, complex problem solving, and topics on leadership and command. The development of training at Levels B and C includes the complex circumstances of detention that are not incident to an armed conflict by a foreign power (for example, peacetime governmental detention and terrorist captivity as a result of operations other than war that require special instructions). Level C training also includes a resistance-training laboratory. The degree of knowledge required by members of the armed forces is dictated by the Service members' susceptibility to capture, the amount of sensitive information possessed by the Service member, and the potential captor's or detaining power's likely assessment of the individual's exploitation value. The Services shall designate personnel to be trained at each level. Current Service regulations prescribe Code-of-Conduct and SERE training requirements and responsibilities. Code-of-Conduct and SERE training enhance an individual's readiness to survive, evade, resist, and escape. They provide an individual with the skills to integrate into established DOD evasion, escape, and recovery systems that support captured and isolated U.S. personnel in conflict and peace. Individuals who have the potential to become isolated require adequate preparations to survive the ordeal. Individual preparation includes training on policy, force structure, and operational concepts to support an individual's recovery and repatriation.

NOTE: The USAJFKSWCS SERE School is the proponent for matters pertaining to SERE and Code-of-Conduct training for SF and the Army.

TRAINING FOR EVASION

3-8. Successful PR is dependent on effective predeployment and premission preparation and training. High-risk-of-capture personnel must be adequately prepared for contingency operations by being thoroughly trained in the following areas:

- Legal status of evaders and PWs (Field Manual [FM] 27-10, *The Law of Land Warfare*, and Chairman, Joint Chiefs of Staff Instruction [CJCSI] 3121.01, *Standing Rules of Engagement for U.S. Forces [SROE]*, provide guidance on the law of land warfare and standing ROE for U.S. forces, respectively).
- Evasion and recovery preparation and available equipment.
- Evasion TTPs, to include initial actions, surface navigation, evasive travel, and camouflage.
- Signaling and communication TTPs, to include radio discipline, codes, team and individual call signs, signaling devices, recovery activation signals (RASs), and GASs.
- Environmental hazards, health risks, and personal protection specific to an area of operation.
- Information on the indigenous population and the captivity environment.
- Conventional recovery and UAR methods.
- The types of PR intelligence data available for their potential area of operation.

CONTINUATION TRAINING

3-9. Evasion continuation training (classroom and field) at the unit level is essential to keep high-risk-of-capture personnel adequately prepared to participate in combat operations. SF soldiers should pursue all opportunities for joint PR training. All high-risk-of-capture personnel should receive survival and evasion field refresher training at least every 3 years and participate in evasion field exercises whenever possible.

PLANNING FOR EVASION

3-10. With an ever-increasing number of SFODAs conducting joint combined exercises for training (JCETs), humanitarian demining operations, and humanitarian operations across the global spectrum, it is imperative that they know how to conduct self-recovery during a civil disturbance or coup or when the sentiment of the HN toward the United States changes. Procedures for recovery must be coordinated with the American Embassy or consulate in each country where an SFODA operates. Doing so may require inclusion in the Embassy's NEO plan, which includes procedures for reaching the Embassy or other safe haven in case of civil unrest. Force protection measures must be in place to ensure a self-recovery plan is in effect and can be executed.

3-11. All personnel subject to isolation in hostile territory must be prepared to evade. Successful evasion is dependent on effective prior planning. (Appendix E provides checklists and voice templates for PR planners.) Evasion planning needs to incorporate, at a minimum, information available from the following sources:

- *Intelligence briefings.* Information on the mission route, enemy defenses and troop dispositions, status of the U.S. and allied military situation, changing attitudes of the enemy populace, and so on is essential when planning potential evasion.
- *SAID.* SAIDs are designed to assist evaders by providing all-source information on an area under hostile conditions.
- *E&E area studies.* E&E area studies are based on operational or contingency planning requirements. These studies meet similar criteria as a SAFE, but differ in that not all SAFE selection criteria can be met because of prevailing political, military, or environmental factors in the region.
- *SERE guides and bulletins.* SERE guides and bulletins provide general information to create a foundation for more specific information (found in evasion charts, SAIDs, and current intelligence briefings) that can be used to build a sound evasion plan.
- *JPRSP.* PR intelligence production is undergoing a profound evolution in product format, presentation, design, and functional production responsibilities. Intelligence support to the PR community in the past has been provided in the form of SAIDs, DARs, E&R studies, regional and country studies, SERE contingency guides, newsletters, bulletins, and so on. These products will be replaced by a single, new collaboratively produced package to support PR—the JPRSP. The JPRSP will be organized in a country or region-wide modular format allowing the planner to "drill down" into the data. The JPRSP is intended to meet the needs of multiple users (PR planners, operators, rescue forces, potential evaders, and so on). In summary, the JPRSP seeks to combine advanced information technology that facilitates collaborative planning in a virtual environment with complementary organizational changes and dynamic processes to improve intelligence support to PR planners, recovery forces, and potential evaders.
- *ISOPREP (DD Form 1833).* The ISOPREP is a DOD form that contains information used by a recovery force to facilitate a recovery. Its use enables the recovery force to authenticate the evader and conclude a successful tactical recovery. The unit and the individual are responsible for ensuring the ISOPREP is properly completed, reviewed, and maintained on file by the SF PR coordinator, and ready to be

transmitted to the appropriate RCC when required. Joint Publication (JP) 3-50.3, *Joint Doctrine for Evasion and Recovery*, Appendix C, provides a sample DD Form 1833.

- *EPA.* The purpose of an EPA is to provide the potential evader with a method to communicate his detailed intentions should he become isolated during the conduct of his assigned mission. The EPA provides a measure of predictability to aid recovery forces and greatly increases an evader's chances of recovery. Personnel must articulate this tentative COA before execution of a mission. The plan is applicable to individuals or small groups. The recovery force is aware that all possibilities cannot be addressed. Since the EPA may be the only common data between the evader and recovery forces, isolated personnel should try to precisely execute their EPA. The EPA must be feasible, supportable by theater recovery resources, properly prepared with the theater SPINS, classified, properly updated, and coordinated with the theater PR architecture. Appendix C discusses EPA planning in detail.

EXTENDED EVASION

3-12. SF personnel should always plan for extended evasion (more than 48 hours and over prolonged distances). Extended evasion differs from short-range evasion in several respects. The evader must be prepared for evasion under both circumstances, because it is impossible to predict how much time it will take or how far an evader may have to travel. Every alternative should be considered before determining a preferred COA. Factors that might be unimportant to a short-range evasion may present major problems during an extended evasion. Considerations in determining a preferred COA include the following:

- The distance from friendly forces may range from hundreds of meters to hundreds of miles. The evader may be discouraged by the knowledge that hundreds of miles of travel over a period of months may be necessary. The will to survive and the knowledge of survival techniques become more vital.

- Personnel control measures such as travel restrictions, security checks, and border crossings may be more prevalent. The evader must anticipate these conditions. The evader should obtain and review information on specific border areas. Knowledge of local customs may provide valuable information to aid evasion.

- The evader should plan for and practice supply economy. The evader should carefully manage items such as shoes, clothing, and supplies to ensure the evader the maximum usage. Isolated personnel in enemy-controlled territory need to decide what equipment to keep and how and where to dispose of the remainder.

- The evader should remember that avoiding capture is important, even if it means leaving the scene of the initial isolation, deviating from the EPA, or leaving valuable equipment behind.

UNPLANNED ASSISTANCE DURING EVASION

3-13. Under some circumstances—especially when seriously injured—the evader may need to seek assistance from the local populace to survive. This should be done only as a last resort. Even when the evader does not require emergency assistance and is doing everything possible to avoid contact with locals, unplanned contacts may occur. All such contacts are very risky, but if handled properly, such contacts could become a source of lifesaving assistance.

SUPPORT TO EVADERS

3-14. When an evader is isolated deep in hostile territory and early recovery is not possible, PR planners may support the evader in two ways. They may support them with—

- *Caches.* Caches may be prepositioned in enemy-controlled territory or in regions subject to being overrun by enemy forces, and their use should be considered in environments where extended evasion is projected. Evaders can use caches as sources of supplies, communications equipment, and other evasion aids.
- *Resupply operations.* When personnel are isolated out of range of rotary-winged aircraft, there are no pre-positioned caches, and recovery forces have been committed to other operations, it may be possible to air-deliver packages to them from fixed-wing aircraft or, in the future, from unmanned aerial vehicles (UAVs). Planners must consider the vulnerability of these resupply aircraft.

EVASION AIDS

3-15. The elements, terrain, hostile military forces, and the local populace challenge evaders. To overcome these challenges and be successful, potential evaders must receive adequate training, information, and equipment before initiating their mission. Whenever possible, potential evaders need to carry evasion aids on their person. Since the incident that causes their isolation in hostile territory may be sudden and unexpected, evaders may be quickly separated from their equipment or may not have time to select those items that would be most useful to them during evasion. Space considerations and clothing configurations may limit the number of evasion aids carried by evaders. As a minimum, potential evaders should carry the following on their person:

- Evasion charts or other maps of the area.
- An extra compass.
- A blood chit.
- A pointee-talkee or other means of communicating with the native populace.
- Fireflies, which are general reference materials to facilitate recall of basic medical, survival, and/or cultural information.
- Items to enhance protection from the elements and provide camouflage.
- A signal mirror.

- Equipment, which permits verbal and visual communications with a recovery force during day and night.

NOTE: References to evasion aids carried on a particular operation should be included in the EPA.

EVC

3-16. The EVC (Figure 3-2, page 3-10) is designed to help isolated personnel evade capture and survive in hostile territory. It also provides evaders with a means of navigating to a SAFE or other recovery area. The EVC program supports operational force requirements with a series of charts that covers geographic areas specifically identified by combatant commanders. The EVC is a derivative of a standard product, the joint operations graphic (JOG), and is made up of about eight 1:250,000-scale JOG charts, usually four on each side. When JOGs of a particular area are not available, Tactical Pilotage Charts (1:500,000-scale) are substituted. The EVC is produced on a very strong waterproof and tear-resistant material. Tailored to cover the individual environmental area concerned, it is a unique, multipurpose product that combines standard navigation charts with evasion and survival information located on the margins. A typical EVC contains localized information on navigation techniques, survival medicine, environmental hazards, personal protection, and water and food procurement, to include color pictures of edible and poisonous plants. The chart is also overprinted with a camouflage pattern similar to the natural ground colors of the area and may aid an evader in hiding when it is used as a shelter or cover. The chart is folded to fit in a flight suit leg pocket and shows an American flag on one of the outer panels. An evader can use this to identify himself, especially when contacting friendly troops in a hostile area. Procedures for ordering EVCs are found in the Defense Mapping Agency (DMA) Map Catalog, Part 1 (Aerospace Products), Volume 1 (Aeronautical Charts and Flight Publications), Section 8 (Special Purpose Products). Although EVCs contain much information, including survival and medical guidance, and can be carried by evaders as they make their way through enemy territory, potential evaders cannot afford to wait until they are already isolated before they study the chart. Effective evasion planning requires that potential evaders be thoroughly familiar with the information on the charts and how to use the charts before departing on their missions.

BLOOD CHITS

3-17. The blood chit (Figure 3-3, page 3-11) is a small sheet of weatherproofed material on which is imprinted an American flag and a statement in English and several other languages spoken by the populace in the operational area. There is a serial number in each corner that identifies each individual chit. The blood chit identifies the bearer as an American and promises a reward to anyone providing assistance to the bearer and/or helping the bearer to return to friendly control. When the blood chit number is presented to American authorities and the claim has been properly validated, it represents an obligation of the U.S. Government to provide compensation to the claimant for services rendered to the evader.

FM 3-05.231

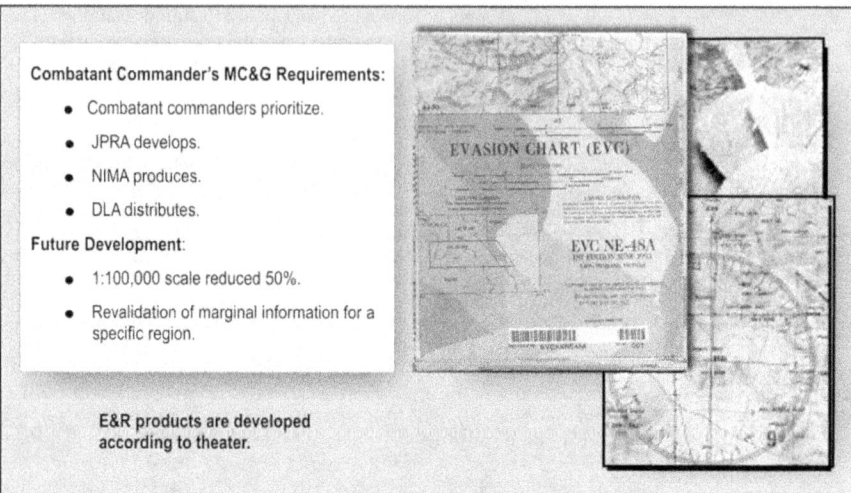

Figure 3-2. EVC

3-18. Although use of the blood chit is at the discretion of the individual to whom it is issued, it should be used only after all other measures of independent evasion and/or escape have failed and assistance is considered vital to survival. Unless the chit is taken by force or threat, individuals should retain it.

3-19. Upon receiving assistance, the evader provides the individual with the blood chit number (either written or one cut from a corner of the chit). If requested, the evader may provide name, social security number (SSN), and signature. Although this additional information could, in the long run, pose a security problem for the Samaritan, he may insist on having more information than just a number to back up his claim for reward. The evader should continually assure the helper that a reward will be provided when the evader safely returns to friendly lines and the number is presented to and properly validated by an official representative of the U.S. Government.

3-20. The blood chit has certain limitations as an evasion aid and form of identification. The individual(s) providing aid may be skeptical of the value of a "number" as something that may produce a reward. The skepticism will increase if enemy propaganda has assured those providing aid of ultimate and certain victory. When the evader tells those providing aid that a reward is conditional upon his (the evader's) safe return to friendly lines, pending validation of the blood chit number by a U.S. Government official, the Samaritan will be forced to choose between the promise of a reward and the known risks in assisting the evader. Also, there is a strong possibility that the person who is about to assist the evader will want to retain the blood chit as tangible evidence to claim the promised reward. When the evader refuses to yield the blood chit, he will certainly arouse suspicion. Overcoming this difficulty will depend largely on the salesmanship of the evader.

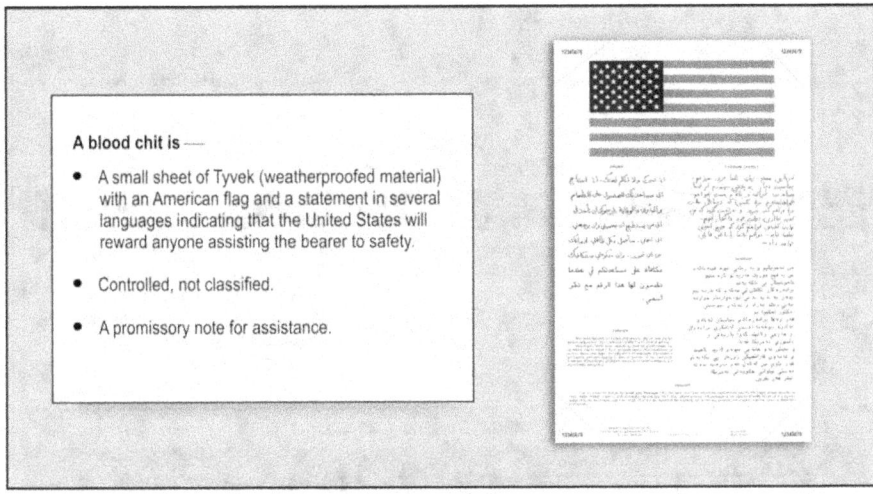

A blood chit is —

- A small sheet of Tyvek (weatherproofed material) with an American flag and a statement in several languages indicating that the United States will reward anyone assisting the bearer to safety.

- Controlled, not classified.

- A promissory note for assistance.

Figure 3-3. Blood Chit

3-21. When the evader is in the hands of friendly guerrilla organizations, use of the blood chit as a means of identification may depend largely on the effectiveness of communications between the guerrilla group and American forces. The evader should expect to encounter some suspicion, because the guerrillas could suspect the chit may have been captured or stolen or could be a skillful counterfeit, and the bearer could be an enemy using it to penetrate the group. In spite of this, the blood chit is a useful evasion aid, and using it wisely provides an excellent chance of return.

POINTEE-TALKEES

3-22. Pointee-talkees (Figure 3-4, page 3-12) are similar to the language guides used by invasion forces during World War II. They contain English phrases on the left side of the page and the same phrases written in the foreign language on the right side of the page. The evader selects the desired English phrase and points to the translation of the phrase beside it. The evader may augment the pointee-talkee by making drawings and signs to help communicate with a local national whose language the evader does not speak or understand. The major limitation of the pointee-talkee, as with the blood chit, is in trying to communicate with illiterates. In many countries, the illiteracy rate can be very high, so personnel may have to resort to pantomime and sign language, tactics that have been relatively effective in the past. Pointee-talkees should be developed in-theater, where the language expertise is available. The JPRA verifies and sanctions the pointee-talkees.

FM 3-05.231

Pointee-talkees —
- Historically developed locally without DOD-wide coordination.
- Are verified and sanctioned at the JPRA homepage.
- Are usually a crisis production (OPSEC).
- Must be translated correctly. The translation is critical and should be done by two native speakers, not a DLI graduate.

SOMALI
INTRODUCTORY MATERIAL (ISBARID)

ENGLISH	SOMALI	SOMALI PHONETIC
I am an American and I need your help, but I do not speak your language. I'll point to the question in your language and you can point to the answer in your language.	Waxian ahayMareykan, waxaanan u baahnahay inaad I caawinto. Laakiin luqaddinna kuma hadlo, Su'aasha oo afkiinna (luqaddiina) ku qoran ayaan farta kuugu fiiqi doonaa cc adna jawaabta oo afkiinna ku qoran farta iigu muuji.	WAH-xahn ah-HAI mah-RAY-kuhn, WAH-xahn-ahn OO BAH-nah-hai IH-nahad ih AH-wihn-toh, lah-KIHN loo-KAIHD-DEE-NAH KOO-mah hahd-LOH. SOO e-SHOH uhf-kee-nah (luh-KAHD-dee-nah) KOO koh-RUHN ah-YAHN FUHR-tuh koo-GOO fee-kuh ay AHD nah juh WUHB tuh oo UHF KEE NAH koo-KOH-RUHN FRHR-tuh ii-goo MOO-chee.

COMMUNICATION (ISLA XIRIIR)

ENGLISH	SOMALI	SOMALI PHONETIC
Will you help me?	Ma I caawini kartaa?	MAI ah-WIH-nee KAHR-tah?
Is there someone that speaks English?	Ma joogas qof af ingiriisi ku hadla?	MEE CHOO-gah KOHF ahfeen-GREE-see kwoh-HAHD-lah?
Are they willing to help me?	Ma doonayaan (rahaan) in ay I caawinaan?	MUH DUH-nay-yuh (RAH-bahn) ibn ay-eh ah-wee-NAHN?

Figure 3-4. Pointee-Talkee Characteristics and Sample Somali Pointee-Talkee

REPORTING OWN SITUATION

3-23. To signal overhead assets, isolated personnel must assist recovery forces by reporting their location via radio either before or during evasion or by nontechnical means such as a coordinated GAS IAW individual's EPA and theater SPINS. National imagery products are often essential in determining an isolated individual's location and condition. Since national imagery assets are controlled from CONUS, tactical commanders do not have the authority to redirect these assets to locate isolated personnel. Once isolated, evaders should—

- Try to report their situation as soon as feasible. However, evaders should consider blending in with the environment, consider the enemy threat in or around their position, and avoid compromise of the security of the contact area. Initial notification will most likely be by radio transmission—either plaintext or secure. The EPA should list alternate communications procedures. Isolated personnel should be prepared to

provide positional assistance to rescue forces to the greatest extent possible.
- Consider known and suspected enemy locations when signaling their position to rescue forces.
- If separated from a group, prepare to provide pertinent information about the dispersal of the group.
- Inform rescue forces if operational developments require alteration of their EPA, thereby adjusting the rescue plan. Isolated personnel should likewise be prepared to receive and follow instructions from the rescue force that require them to alter their EPA to adapt to the current situation.

SIGNALING

3-24. Signaling between isolated personnel and rescue forces may be severely limited or restricted by the capabilities of signaling devices, terrain, weather, medical status, and enemy activity. Isolated personnel should pay close attention to and explicitly follow the instructions of rescue forces to the maximum extent possible given the tactical situation. Isolated personnel should be prepared to respond quickly and accurately to authentication procedures and requests for ISOPREP information. The most important action isolated personnel can take to assist their recovery within denied territory is to provide evidence to friendly forces that they are alive and still evading capture. Usually, the best time to provide this evidence is at initial notification. Afterward, providing such evidence should be done more covertly and should consider the enemy situation. Evaders should also consider the following:

- All personnel should be able to properly use all issued signaling devices. They should become familiar with these devices before they are needed.
- All personnel must be able to improvise signals to improve their chances of being sighted. For instance, a signal mirror can be made from any shiny metal or glass.
- Isolated personnel should carefully select a signaling site that enhances the signal. For instance, they could move up and out of valleys, drainage, and canyons before transmitting.
- The signal site should have natural or man-made material readily available for immediate use. For example, high ground or an open space that contains material that could be used for passive signals such as visual GASs (Figure 3-5, page 3-14).
- Isolated personnel should take particular care when signaling to avoid disclosing their position to the enemy. They should use the terrain to mask transmissions from the enemy.

- Batteries are not required.
- Site selected should be unobservable to ground forces and away from shadows.
- Construction principles:
  - Size: 18 feet x 3 feet.
  - Ratio: 6:1.
  - Letter or shape.
  - Contrast.
  - Angularity.
- Materials:
  - Natural.
  - Man-made.
- Quality construction is key.

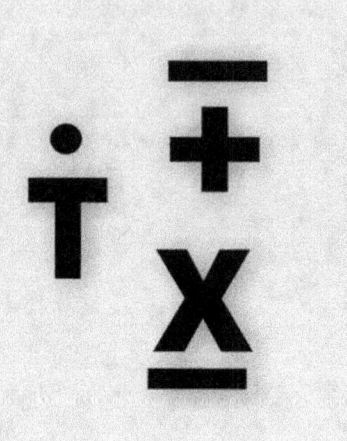

Figure 3-5. Ground-to-Air Signals

CONTACT PROCEDURES

3-25. When PR planners develop standards and guidelines for contact procedures, several factors should be considered. These include, but are not limited to, the following:

- The type of terrain in the operational or recovery area (desert, jungle, urban, and so on).
- The equipment and evasion aids that may or may not be available to isolated personnel.
- Enemy capabilities (such as air superiority, reconnaissance and/or direction-finding capabilities, and other enemy capabilities).
- The availability and training level of recovery forces.

3-26. To make contact between them possible, evaders and recovery forces employ either technical or nontechnical communications, or both. Theater CSAR SPINS and individual EPAs must include enough guidance to allow contact to occur in situations beyond the capability of conventional forces. Published recovery area data should include contact points. Contact procedures should be IAW existing joint TTP to preclude complications. Although no template can be used for every situation, PR planners should consider the following basic information in all planning documents:

- *Signal plan.* Evaders should be prepared to emplace two types of signals: RAS and load signal.
- *Contact plan.* PR planners preplan the procedures to facilitate the linkup between the evader and the recovery force.

INDIVIDUAL RESPONSIBILITIES

3-27. Isolated personnel have a responsibility to assist in their own rescue to the maximum extent possible. Individual responsibilities of isolated personnel and evaders include the following:

- Completing and reviewing their ISOPREP and EPA before a mission.
- Knowing and complying with theater-specific CSAR SPINS.
- Thoroughly understanding CSAR authentication, notification, and reporting requirements.
- Being familiar with survival techniques and equipment.
- Being familiar with CSAR and NAR operations to assist in their own recovery.
- Being mentally and physically prepared to survive and evade for indefinite periods.

3-28. Because CSAR assets are extremely vulnerable during the recovery phase, the evader will not normally be recovered until his identity has been verified. The evader can expect to provide the recovery force with authentication data contained in his ISOPREP and theater SPINS code words or visual signals contained in the theater SPINS or in the individual's EPA. The evader should be prepared to use all signaling devices IAW CSAR force instructions. If recovered by a helicopter, the evader should turn away from the landing helicopter to avoid flying debris and hold his position until contacted by the contact element. The evader should carefully follow all instructions given by the recovery force. The evader will be authenticated and escorted to the helicopter by the contact element.

3-29. If recovered by a NAR element, the evader should be quiet, avoid resisting, and carefully follow all instructions to avoid compromising the security of the recovery force. All of the evader's operational equipment, to include weapons, munitions, technical and nontechnical communications equipment, and all other potentially compromising material, will be separated from the evader. Whenever possible, the evader's equipment will be safeguarded and exfiltrated with the evader to friendly control. The evader will be briefed on what to expect during the remainder of the recovery operation. These actions protect the recovery force, pending authentication of the evader, and also provide security from compromise for the recovery force and the evader. Regardless of his status, the evader should understand that the recovery force commander is the mission commander until the mission is complete. Evaders must understand that being recovered does not equate a quick return to friendly control. There may be occasions when the recovery force that made the initial contact with the evader cannot, for operational reasons, deliver the evader safely to friendly territory. In such cases, the evader may be turned over to another NAR element or to a conventional force to complete the extraction from hostile territory. In all phases of the recovery operation, the evader receives recontact instructions in case he becomes separated from the recovery force. The evader can best assist in the recovery process by doing exactly as directed by the recovery force.

SECTION II – OPPORTUNE SUPPORT TO PR

3-30. SF possesses the skills, capabilities, and modes of employment to provide support to PR. SF deploys in small teams trained to operate clandestinely in enemy territory or denied areas. These capabilities make it an effective PR asset in situations where an SFODA is already present near the PR requirement. An SFODA may be diverted from its primary mission to recover an evader, or it can conduct a recovery operation as follow-on to its primary mission.

DIVERSION FROM PRIMARY MISSION IN SUPPORT OF PR

3-31. SF units conducting other operations may be redirected by the JFC, through the JFSOCC, to provide immediate assistance to PR. SF in this capacity is normally executing deliberate missions in support of the JFC's campaign plan. The value of the primary target must be weighed against the recovery mission. A risk analysis is part of this decision process. Redirecting infiltrated SFODAs to perform recovery operations will certainly affect performance of the original SO mission. Therefore, the JFC, in coordination with the JFSOCC and with input from the JSRC, makes the decision to redirect. Figure 3-6, page 3-17, illustrates an SFODA being diverted from an SO to recover an evader.

3-32. In reaching his decision to redirect an SFODA, the commander must consider the geographic location and physical condition of the isolated personnel, the immediate risk to the evader(s), and the necessary freedom of movement that allows the SFODA to contact them. Mission planners will use all available information, including observation and fields of fire, avenues of approach, key terrain, obstacles, and cover and concealment (OAKOC) and mission, enemy, terrain and weather, troops and support available—time available, and civil considerations (METT-TC) to formulate the staff recommendation for the commander's decision to redirect the SFODA. Once planners have established the need to redirect and the JFSOCC has approved the CONOPS, the SFODA will be notified via established communications channels using standardized message formatting found within the current SAV SER SUP. Any one mission tasked to a deployed detachment could require numerous messages transmitted from the SFOB to the detachment. For example, an SFODA may receive NICKY, QUART, SAFER, and GRAZE reports to provide enough information to allow them to plan for a single redirected mission.

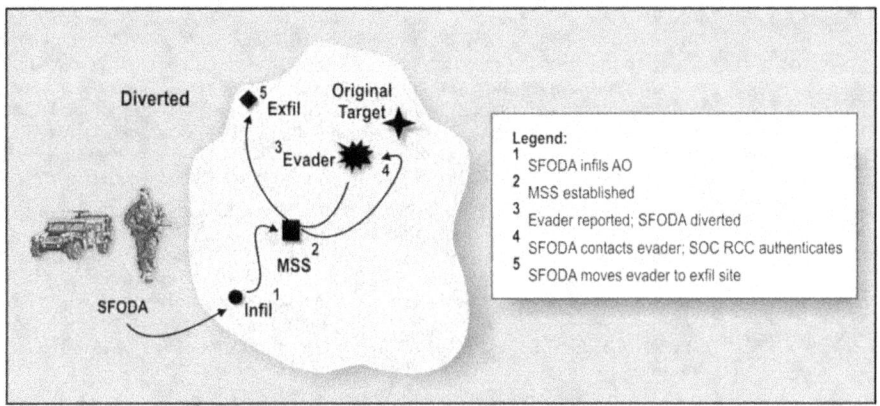

Figure 3-6. SFODA Diverted in Support of PR

3-33. Once in the recovery area, the SFODA will establish contact with the evader following information provided in reports received. This information is gleaned from the evader's EPA, the daily ATO SPINS, and the evader's ISOPREP. Coordination to assist the SFODA and evader in establishing linkup is done through the JSRC and appropriate RCCs. If the isolated personnel are not contacted at either the primary contact point or the alternate contact point, the SFODA will return to the mission support site (MSS) and wait for the isolated personnel to reestablish contact with the JSRC.

3-34. The SFODA can perform a limited ground search. This is only done while the detachment is doing an initial search for signals emplaced by the isolated personnel. Any searches conducted place the detachment and isolated personnel at increased risk of capture or discovery by enemy forces.

MULTIPLE AND FOLLOW-ON MISSIONS

3-35. Operational requirements often cross over doctrinal mission lines. An SFODA can conduct an operation involving multiple missions. Multiple missions are appropriate when operational requirements demand the application of TTP drawn from more than one doctrinal mission. For example, an SFODA can conduct special reconnaissance (SR) as its primary mission, with a be-prepared-to mission of providing support to PR.

3-36. During the mission planning process, SFODAs may be assigned a follow-on mission to be conducted before their exfiltration from an AO. For example, an SFODA can conduct a mission against a specific target and then link up with a resistance organization to conduct UAR. Preplanned multiple missions are appropriate when the operational SFODA can be supported and resupplied within the AO.

3-37. Because of their inherent skills and vast access to deep battlespace, SFODAs may also receive follow-on missions while performing their initial mission. For example, an SFODA conducting another mission may be directed to recover a downed aircrew or other designated personnel. Such follow-on missions are often tasked to the SFODA because of their access and abilities. Planning, rehearsing, and preparing for additional operations even while deployed is an invaluable SF trait. Follow-on missions generally rely on extensive coordination and support from the FOB and appropriate RCC (when involved) to guarantee mission success. SFODAs may receive follow-on missions when the importance of that mission justifies the additional risk.

3-38. SF units conducting SO (for example, DA, SR, or UW) may be given the mission of recovering isolated personnel following completion of their primary mission. These recovery operations can only be conducted if the enemy situation permits and resupply operations, if required, can be conducted. In many instances, the SFODA will exfiltrate the area with the isolated personnel upon completion of the mission. Figure 3-7 shows an SFODA conducting a PR operation after its primary SO mission was conducted.

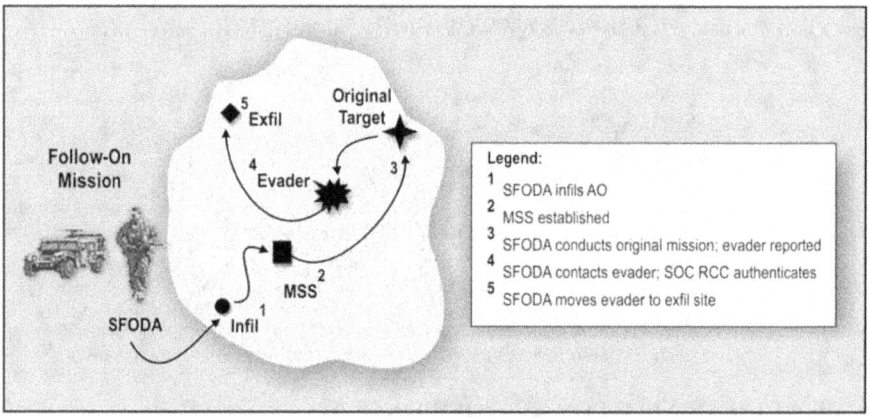

Figure 3-7. SFODA Follow-On Mission in Support of PR

SECTION III – UNILATERAL RECOVERY

3-39. CSAR, whether conducted by a single component or a CSARTF, occurs because of a PR incident. A unilateral CSAR results from the responsibility of units to perform CSAR in support of their personnel who become isolated.

UNILATERAL RECOVERY RESPONSIBILITY

3-40. USSOCOM, with its Service-like responsibility, must conduct recovery of its own forces when possible. All USSOCOM subordinate forces, including SF, are also responsible for performing CSAR in support of their own operations and consistent with their assigned functions, organic capabilities, and limitations.

UNILATERAL CSAR

3-41. When SF conducts a unilateral CSAR, it is a planned combat operation where SF elements self-deploy into hostile territory and travel via organic transportation (for example, high mobility multipurpose wheeled vehicle [HMMWV], dismounted, or small boats) to a predetermined rendezvous point to make contact with an evader. Once contact has been made, the recovery force and the evader can then exfiltrate back to friendly control (Figure 3-8).

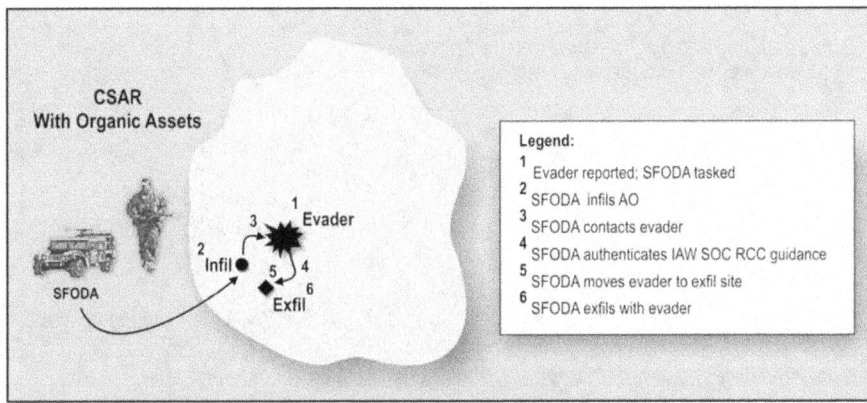

Figure 3-8. SF Unilateral CSAR

SECTION IV – JOINT RECOVERY OPERATIONS

3-42. The JFC, using his command authority through the JSRC, may task the JFSOCC to provide direct support to a CSARTF. SF units, as part of a CSARTF, may be required to conduct JCSAR operations as a collateral mission. Normally, SF is not manned, trained, or equipped for collateral activities. The role of SF in support of JCSAR will generally fall into one of the following categories: SF as the security and contact force for a CSAR special operations aviation regiment (SOAR) platform, exfiltration of an evader recovered by an unconventional assisted recovery team (UART), and exfiltration of an evader recovered by an RM.

SF AS THE SECURITY AND CONTACT FORCE FOR A CSARTF

3-43. When in direct support of a JCSAR operation, SF units can conduct a CSAR operation, in which an SFODA can be inserted into hostile territory, make contact with an evader, and exfiltrate with the evader back to friendly control (Figure 3-9). This type of operation is most common when the detachment is providing security for and serving as the contact element for an ARSOF platform. It can also occur when a contact team is required to move from the HLZ of the recovery helicopter to a contact point with an evader.

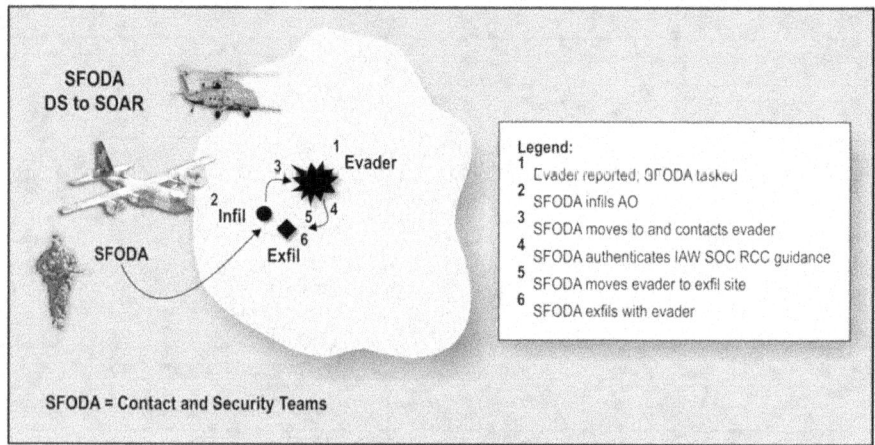

Figure 3-9. SFODA in DS of SOAR CSARTF

3-44. Depending on the situation, an SF element can be inserted into hostile territory, travel overland to a prebriefed or predetermined rendezvous point, and make contact with an evader. Once contact has been made, the SFODA and the evader can travel to an exfiltration point and be extracted via conventional or SOF aircraft back to friendly control. Figure 3-10, page 3-21, shows an SFODA in DS to a CSARTF.

3-45. The JFC may task SF units to provide DS to the AFSOC as a security element during JCSAR operations. While conducting these operations, the SFODA will be placed under the OPCON of the airborne mission commander. In this role, the detachment will provide security to the helicopter and crew and assist the AFSOC pararescue personnel (PJ) and combat controllers (combat control team [CCT]) in recovering the evader. This is not a core mission of SF and is normally associated with a rescue force on strip alert. Although not a primary mission, SF units have the latitude to perform the mission as a collateral SO activity.

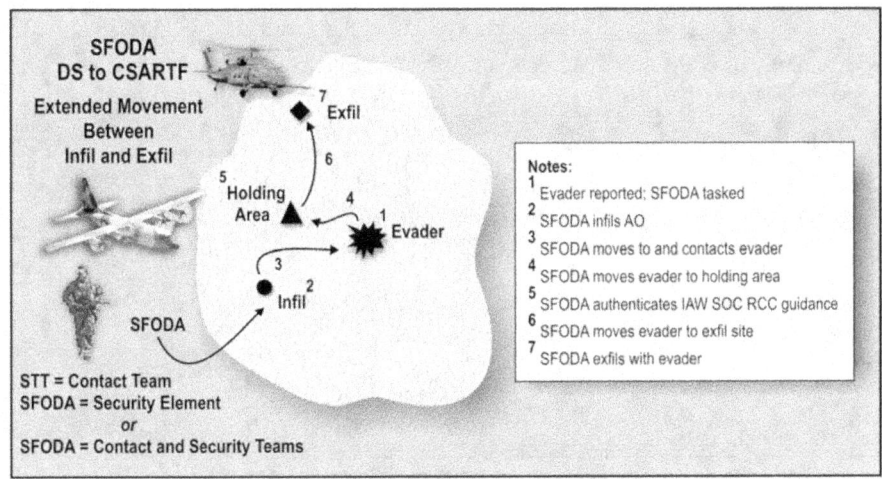

Figure 3-10. SFODA in DS to CSARTF

EXFILTRATION OF EVADER RECOVERED BY A UART

3-46. SF units in DS of a CSARTF may be required to exfiltrate an evader recovered by a UART. The area of access of a UART may be limited in scope. The UART may have the ability to contact, authenticate, support, and move the evader only within its area of access. Operational limitations or logistical constraints make it preferable to exfiltrate the evader back to friendly control at the earliest opportunity to allow the UART to continue with its assigned mission. As shown in Figure 3-11, page 3-22, the SFODA can infiltrate hostile territory, receive the evader from a UART at a crossover point, authenticate the evader, and exfiltrate with the evader back to friendly control. The crossover of the evader between the UART and the recovery force would require detailed coordination between the UARCC, the SOC RCC, and the JSRC. This type of operation does not necessarily require direct contact between the UART and the recovery force at the crossover point. From the perspective of the SFODA, this is simply a conventional CSAR mission where the rescue force plans for and moves to a known location to pick up an evader.

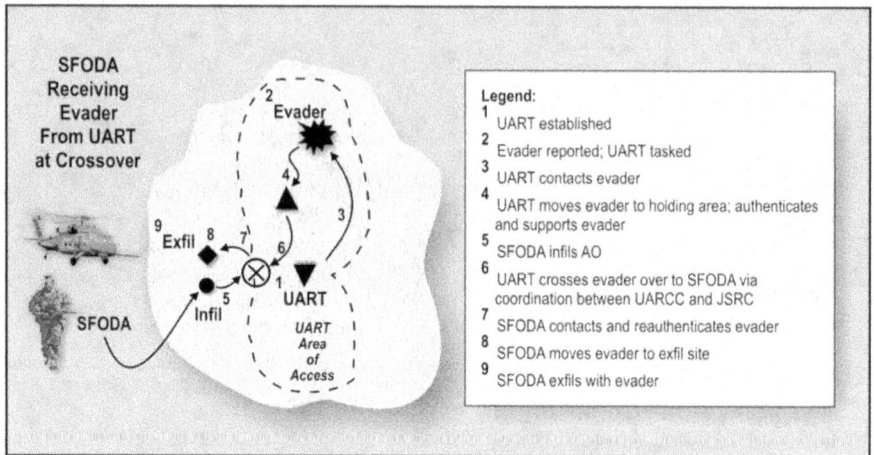

Figure 3-11. Crossover From UART to an SFODA in DS of CSARTF

EXFILTRATION OF AN EVADER RECOVERED BY AN RM

3-47. SF units in DS of a CSARTF may be required to indirectly complement NAR operations. For example, an RM may have the ability to contact, authenticate, and support an evader, but only move the evader to an area of decreased threat accessible by a CSARTF. The RM may have a limited area of access, which precludes it from exfiltrating the evader to friendly control. As shown in Figure 3-12, page 3-23, the SFODA may insert (as part of a CSARTF) into hostile territory, service a crossover point, and make contact with the evader. The SFODA then authenticates and moves the evader to an exfiltration site located outside the RM's area of access. The SFODA and evader then exfiltrate via SOF or other conventional recovery platform back to friendly control. The crossover of the evader between the RM and the SFODA would require detailed coordination between the UARCC, the SOC RCC, and the JSRC. These recovery missions are conventional CSAR operations, and the inclusion of the UARCC in the planning effort is coordinated through the JSRC. The SFODA does not require direct contact with the RM.

FM 3-05.231

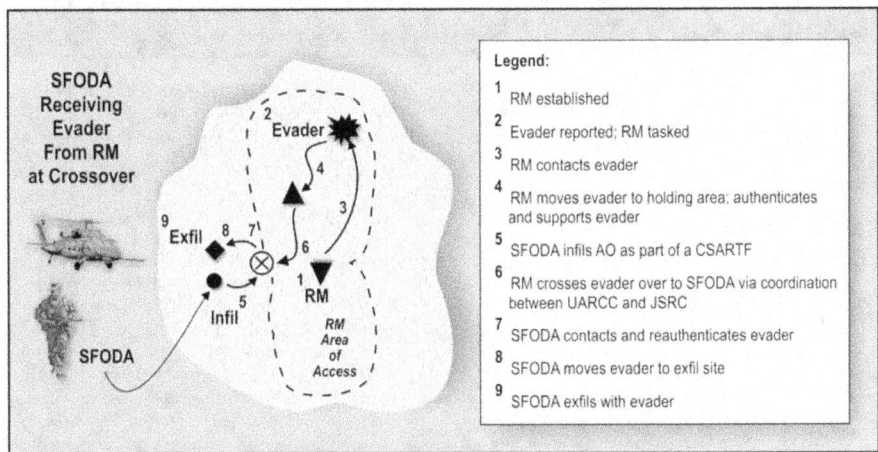

Figure 3-12. Crossover From RM to an SFODA in DS of CSARTF

OPERATION ALLIED FORCE

3-48. A vignette of operation ALLIED FORCE is provided below. The SOF CSAR teams successfully rescued downed pilots.

NATO initiated Operation ALLIED FORCE on 24 March 1999 to put an end to Serbia's violent repression of ethnic Albanians in Kosovo. The 19-nation ALLIED FORCE coalition conducted an unrelenting bombing campaign in Serbia and Kosovo for 78 days, eventually forcing Serbian President Slobodan Milosevic to withdraw his forces from the province and stop the "ethnic cleansing" of Kosovar Albanians. The bombing strategy did not prevent Serbia from forcing an estimated 800,000 refugees out of the country, however, which produced an enormous humanitarian crisis in the neighboring states of Albania and Macedonia. Furthermore, the air campaign did not eliminate all of Serbia's surface-to-air missiles, which managed to shoot down two U.S. aircraft.

The Joint Special Operations Task Force—Noble Anvil (JSOTF-NA) joint special operations air component commander (JSOACC) had the primary responsibility for CSAR planning and execution during Operation ALLIED FORCE. SOF successfully rescued the only two U.S. pilots downed during ALLIED FORCE. SF units integrated into the CSAR mission as the ground force provider. In separate missions that each took less than a minute on the ground, SOF CSAR teams, which were comprised of PJ, CCT, and SF security elements, rescued an F-117A pilot who was shot down near Belgrade on 27 March and an F-16 pilot shot down in western Serbia on 2 May. On each occasion, a mixture of MH-53 Pavelow and MH-60 Pave Hawk helicopters were used to retrieve the downed fliers. These rescues had profound effects on the outcome of the operation.

3-23

SECTION V – SUPPORT TO MULTINATIONAL PR OPERATIONS

3-49. Throughout the history of SF, SF personnel have participated in multinational and combined operations. SF personnel currently serve in joint combined operations and coalition support teams as advisors to multinational or coalition forces, providing much expertise in the conduct of operations.Sf support role

SF SUPPORT ROLE

3-50. Support to PR operations consisting of planning and advice are commonplace. It is not as common to see U.S. forces employed in the role of security on an allied aircraft or under the command of an allied commander. Conventional PR operations in support of U.S. personnel will rarely employ allied or coalition C2. Communications and technical equipment compatible with PR operations are endemic to U.S. forces. The potential role of SF in the multinational PR arena exists; however, issues involving release of sensitive TTP and current U.S. policy placing the C2 of U.S. forces under a U.S. commander make it unlikely. The role of SF is to train, advise, and assist. There is no tactical application of SF support to coalition PR. Although the limited potential exists for SF to support coalition operations, it is more likely that SF will be involved with coalition UAR.

UAR IN THE COALITION ENVIRONMENT

3-51. UW includes coalition and combined forces. The sensitive nature of UAR TTP presents OPSEC considerations and C2 challenges for combined forces and adds operational uncertainty during those phases directly controlled by foreign SOF. Despite these challenges, coalition UAR offers greater flexibility in planning, provides the warfighter with more options for his overall PR plan, and in many cases, offers the only potential recovery force within the area of operation.

COMMAND AND CONTROL OF COALITION FORCES

3-52. The C2 structure or method used by our allies will not affect NAR mission management. The key to success is coordination. The UARCC director or his representative will directly coordinate with his counterpart from the allied UARCC or its equivalent. U.S. commanders retain control of their respective forces; allied or coalition commanders retain control of their respective forces. Effective coordination will facilitate the transfer of only that information required to perform a specific act in which that force has the capability or sufficient information to effect the linkup or crossover of an evader to U.S. forces.

COORDINATION FOR COMBINED OPERATIONS

3-53. When coalition forces have access or the capability required to support a PR event, the tactical planner of each nation's UARCC, or equivalent, will coordinate with the coalition forces. The tactical planner will coordinate the initial contact with the evader and the subsequent planning for linkup or crossover procedures. In many circumstances, U.S. evaders or isolated personnel will have the means to communicate with U.S. entities only. If the only forces with the capability to support U.S. evaders or isolated personnel are allied or coalition NAR forces, the coordination between allied or coalition tactical planners and the tactical planners in the UARCC will be the key to a successful recovery. Once an evader is in the control of coalition NAR forces, coordination between the participating UARCCs will be the only available means to conduct initial authentication of the isolated person.

DOCTRINAL TTP FOR COALITON UAR

3-54. Doctrinal TTP for coalition UAR should not be developed because of the vast differences in methods of operation used by other nations' SOF. The need for nations to protect their own TTP makes coordination of specific acts a preferred method. As such, isolated personnel must be reauthenticated once they are moved from the control of another nation's forces to U.S. control. As with PR events involving only U.S. forces, the evader must follow the instructions of the recovery force, regardless of their nationality or status.

JPRA J-32 PERSONNEL

3-55. A team that includes JPRA J-32 personnel will conduct debriefings of isolated personnel who have been recovered and repatriated with the assistance of NAR forces. Any debriefing in which a returned individual indicates they were recovered or assisted by unconventional means must be stopped until representatives of the authorized DOD executive agency (JPRA J-32) are present.

SECTION VI – UNCONVENTIONAL ASSISTED RECOVERY

3-56. The greatest contribution of SF units to a combatant commander's PR architecture is their inherent capability to conduct UAR. Title 10, USC, specifies that all unconventional operations are within the purview of SOF. Therefore, the term "NAR" was established to incorporate the participation and contributions of other government agencies and indigenous or surrogate forces into PR operations conducted beyond the capabilities of conventional recovery forces. NAR is the recovery of isolated persons by SOF UW ground and maritime forces and OGAs specially trained to develop NAR infrastructure and interface with or employ indigenous or surrogate personnel. These forces operate in uncertain or hostile areas where CSAR capability is infeasible, is inaccessible, or does not exist to contact, authenticate, support, move, and exfiltrate isolated personnel back to friendly control. NAR forces generally deploy into their assigned areas before strike operations and provide the JFC with coordinated PR capability for as long as the forces remain viable. UAR is NAR conducted by SOF. As a subset of UW, SOF conduct UAR unilaterally with indigenous or surrogate personnel or OGAs IAW Title 10 or Title 50, USC, employing compartmented TTP. UW is the foundational mission of SF. UW encompasses a broad spectrum of military and paramilitary operations conducted in enemy-held, enemy-controlled, or politically sensitive territory. The military aspects of UAR are classic UW for which SF units are specifically organized, trained, and equipped. In the conduct of UAR, SF units are normally deployed into a JSOA before strike operations to support recovery operations. SF personnel conduct a thorough analysis of the area before insertion. SF units may be directed to develop a UARM or provide support to an RM established by OGAs. Generic capabilities required for UAR include an understanding of UW theory and insurgent tactics, knowledge of clandestine operations, area and cultural orientation, language proficiency, small-unit tactical skills, advanced medical skills, and communication skills.

3-57. SF units may be employed in a clandestine manner to support PR. SF units can assist PR operations before or during hostilities while operating within a specified JSOA. Incorporation of indigenous or surrogate forces that are directed or supported by SF units offers the JFC additional flexibility for PR. The theater SOC is normally tasked by the JFC to plan for and execute UAR in coordination with the JSRC in support of the theater PR plan. The intent of UAR is to bring isolated personnel into contact with, and ultimately into the custody of, a recovery force as soon as possible and then move the isolated personnel to an area where exfiltration to definitive U.S. Government control can occur. When properly tasked, SF units may provide a recovery force that can be pre-positioned in their AO before strike operations to conduct UAR operations. SF accomplishes the UAR mission by employing UARTs that may operate unilaterally, with OGAs, or with indigenous or surrogate forces.

SPECTRUM OF UAR OPERATIONS

3-58. The range of employment options for a UART is limited only by the operational situation. UAR operations take place in enemy-held territory or denied areas and support overall theater PR operations. UAR missions are conducted in those circumstances beyond the capabilities of conventional forces and in which SOF may act with indigenous or surrogate personnel, other elements of DOD, OGAs, or multinational forces to effect PR. Unlike CSAR, which is reactive in nature, UAR is proactive in anticipation of PR events. Additionally, UAR operations differ from conventional PR operations in the—

- Degree of political risk.
- Operational techniques.
- Relative independence from friendly support.
- Dependence on detailed operational intelligence.
- Potential use of indigenous or surrogate forces.

3-59. UAR operations differ from intelligence-related activities in their operational focus. UAR operations require specialized TTP executed by small, specially trained and configured organizations. These organizations must be capable of independent operations where the use of conventional forces is inappropriate or infeasible or where conventional forces are nonexistent.

SPECIFIED TASKS

3-60. PR is an integral part of military operations. Combatant commanders are responsible for developing plans and requirements to report, locate, support, recover, and repatriate or return isolated personnel. UAR focuses primarily on the "recovery" task of a PR operation. UAR relies on the UW capability of SF to conduct five specified tasks. These tasks remain constant, regardless of the mission profile (for example, UAR conducted unilaterally by a UART or UAR conducted in conjunction with an RM). The tasks are separate but may be conducted concurrently or sequentially. However, all tasks must be conducted to complete the mission. The five specified tasks of NAR or UAR are:

- Contact isolated personnel.
- Authenticate isolated personnel.
- Support isolated personnel.
- Move isolated personnel.
- Exfiltrate isolated personnel to friendly control.

Though the above-listed tasks remain constant, the TTP employed to accomplish them may be many and varied, allowing for particular unit or individual training strengths, equipment, or employment criteria.

3-61. During the recovery process, recovery personnel consider contingencies in case the evader becomes separated from the recovery force. These contingencies will facilitate recontact with the evader later IAW the capabilities and limitations of the recovery force. During all phases of the recovery process, recovery force personnel should brief the isolated personnel on pertinent procedures, restrictions, and recontact plans while the isolated person is in the custody of the UAR recovery force.

CONTACT

3-62. Contact entails all actions that lead to the positive control of isolated personnel. These actions may include locating the isolated personnel, use of technical and nontechnical communications, and employment of various conventional and unconventional PR techniques and procedures. Locating isolated personnel is the most critical step in the process to return them to friendly control. The ability to rapidly and precisely locate isolated personnel requires the integrated and synchronized capabilities of all recovery assets from the national level to the component level.

3-63. This process begins with individual training in evasion procedures, service and unit training in PR planning and operations and evasion procedures, theater PR and evasion procedures, national procedures, and joint TTP. UART TTP employed to locate isolated personnel must maximize and exploit the opportunities for success. TTP must be tailored to METT-TC considerations, UART training and equipment, and the applicable recovery considerations. UART recovery considerations include—

- Availability of resources, capabilities, and limitations.
- Task-organizing.
- Recovery criteria.
- Location and physical condition of the evader.
- Access.

- Time.
- Movement.
- Capacity.
- Risk assessment.

A UART must be able to locate isolated personnel in all weather and light conditions, and in both rural and urban environments.

Contact Considerations

3-64. Whenever an individual is recovered with the assistance of a UAR force, the most critical aspect of the recovery is the moment the isolated personnel and the recovery force first meet. This period is very dangerous, because it requires two parties, unknown to each other and located in hostile territory, to meet without being detected by either enemy forces or elements of the local population and without compromising either party's security. Contact between isolated personnel and a recovery force requires comprehensive preplanning. The JSRC, assisted by component intelligence and operations specialists, must ensure the appropriate contact procedures are developed and provided in the CSAR SPINS. The SPINS impart detailed evasion and recovery procedures primarily for potential isolated personnel. The JFC must ensure that joint force components are familiar and comply with CSAR SPINS procedures to preclude placing recovery forces at great risk and to avoid significant recovery delays or capture by enemy forces.

NOTE: Publications on advanced special operations techniques (ASOT) include specific and detailed TTP.

Coordinating Contact

3-65. Recovery planners must ensure the procedures for contact are as simple as possible, yet still afford the requisite security measures essential to the protection of the UAR force and the isolated personnel. The isolated individual must make a conscious decision to deviate from his normal evasion routine to initiate actions that will signal his intent to make contact with a UAR force. The isolated individual, now an evader, is no longer avoiding all unknown persons; he is now looking for someone who will perform a specific act indicating his intent to assist. There are two basic scenarios for coordinating contact:

- *Technical communications with isolated personnel.* The isolated individual has established technical communications (for example, radio communications [Figure 3-13, page 3-29]) with friendly forces. The isolated individual could receive contact instructions via radio from an outside source, such as an overhead Airborne Warning and Control System (AWACS) or the JSRC. The UAR force will be alerted by the UARCC to service a coordinated contact point. All contact procedures are coordinated at the JSRC and UARCC and then passed to the isolated individual and the UAR force. For instance, if an isolated individual is to make contact at a location known to the UAR force, the isolated individual can be directed to that location and make contact IAW the prearranged contact procedures received by radio. The isolated individual would know when to expect contact, and the UAR force would know the evader is in place before initiating contact procedures. Tactical communications with the isolated personnel affords the UAR force greater flexibility and greatly increases its ability to make contact with an evader.

- *Absence of technical communications with isolated personnel.* The isolated individual has not established technical communications. This situation requires that the isolated individual include potential contact procedures with a recovery force without communications in his EPA (Figure 3-14, page 3-30). For instance, the isolated individual could state in his EPA that he will initiate signals that would assist a recovery force in finding his location and making contact. (Isolated individual states in his EPA that he will put out a recovery activation signal and will be located 100 yards due north from that location.) The isolated individual must consider day and night signals. The CSAR SPINS serve as a comprehensive guide providing the signals an isolated person must consider to facilitate his recovery.

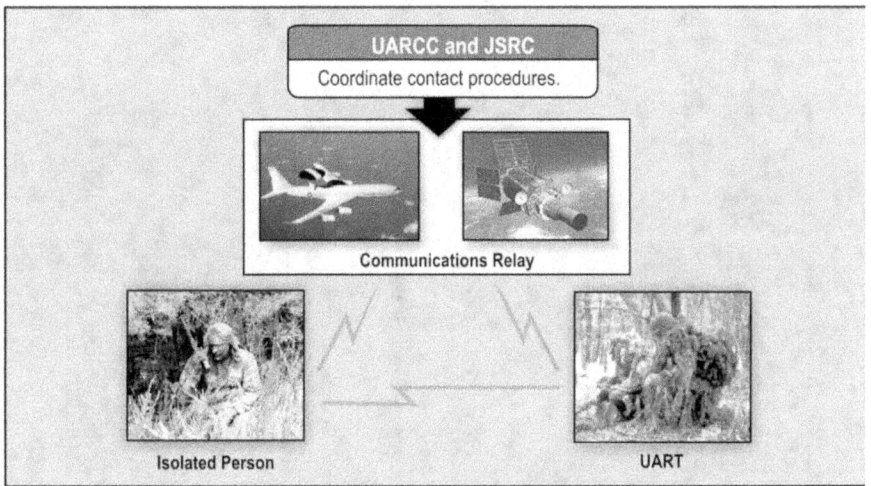

Figure 3-13. Contact With Communications

3-66. Successful contact is the last event in a series of specific actions isolated personnel must initiate, and the entire sequence is based on the guidelines and policies coordinated by the JSRC. Once contact is established, control of the isolated individual must be maintained throughout the operation.

FM 3-05.231

Developing Standards and Guidelines for Contact Procedures

3-67. When recovery planners develop standards and guidelines for contact procedures, they consider several factors. These include, but are not limited to, the following:

- Type of terrain in the operational or recovery area (desert, jungle, urban, and other types of terrain).
- Equipment and evasion aids that may or may not be available to isolated personnel.
- Enemy capabilities, such as air superiority and reconnaissance or direction-finding capabilities. The JSRC coordinates and disseminates the theater-specific policies that will guide isolated personnel actions from the moment of isolation to the actual contact with a recovery force. The JSRC should widely distribute these policies to ensure commanders, potential isolated personnel, recovery forces, and mission planners understand their respective roles in the operation.

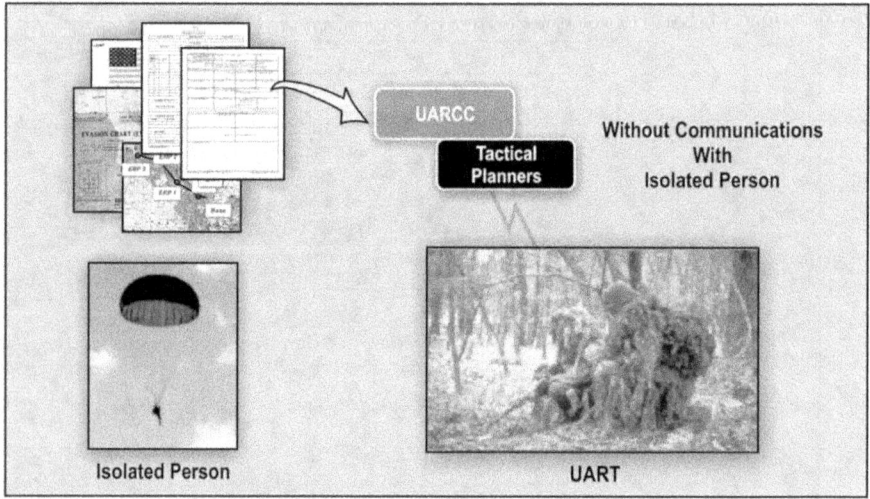

Figure 3-14. Contact Without Communications

3-68. The unit PR coordinator is responsible for ensuring potential isolated personnel provide a comprehensive EPA that complies with theater policies, SPINS, plans, and orders. Isolated personnel will follow their EPA as closely as possible. Their degree of adherence to and skill in execution of EPA activities are crucial to successful contact and recovery operations.

AUTHENTICATE

3-69. As soon as tactically feasible, the UAR force will initiate the authentication process to determine that the isolated individual is in fact the person it has been tasked to recover. Authentication is, "in evasion and recovery operations, the process whereby the identity of an isolated person is confirmed" (JP 1-02, *Department of Defense Dictionary of Military and Associated Terms*). Once an isolated individual has been located, contacted, and taken into the custody of the UAR force, the isolated individual must be authenticated. Until the authentication is complete, the individual is safeguarded not only from the enemy but also to protect the operational security of the UAR force. Although the level of control exerted over the isolated individual will be much more stringent before authentication, it should be proportional to the situation. Recovery personnel should use sensory deprivation and physical restraints only if the perceived level of risk is commensurate with these measures and not as a matter of SOP. Once the isolated individual is authenticated, recovery personnel may relax the control measures exerted over him. However, NAR forces must maintain control throughout the isolated individual's association with NAR forces. The isolated individual could be suffering a significant amount of physical, emotional, and mental stress; may be suffering from some form of exposure and/or deprivation; and may have sustained injuries during or after the isolating event. Thus, safeguarding the isolated individual also means protecting him from his own impaired judgment and decision-making abilities.

3-70. Within a theater of operations, the JSRC manages authentication. The JSRC establishes the thresholds for authentication. In the conduct of UAR operations, the UAR force facilitates the process. The UARCC is the linkage between the two. The JSRC forwards the isolated individual's ISOPREP and EPA information to the UARCC. The UARCC forwards only the minimum amount of information, usually in the form of questions, required to perform authentication. Definitive authentication may be based on the information in the theater CSAR SPINS, isolated individual's ISOPREP card or EPA, detailed physical description, digital photo, fingerprint information, or any combination thereof. The UAR force conveys to the UARCC, the isolated individual's responses to the authentication questions. The JFSOCC has authentication authority, which he may delegate to the UARCC director. The UARCC confirms or denies authentication. The UAR force or parts thereof may only have access to nontechnical, clandestine communications. The process of authentication of the isolated individual may take days or weeks. As such, recovery planners should consider this potential time lag when considering control and support of the isolated personnel.

SUPPORT

3-71. Support includes all actions taken to provide sustainment to the isolated individual and ensure his well-being. The isolated individual may not be in the best physical, mental, or emotional condition upon making contact with the UAR force. The UAR force should provide the greatest degree of support possible without compromising operational security. The UAR provides the following types of support:

FM 3-05.231

- *Sustain.* When possible, isolated personnel should be provided sufficient nourishment, clothing, shelter, safeguarding, and medical care to restore and sustain their health and physical condition. Evading may have drained them of strength and energy that must be restored before subsequent movement.

- *Monitor and assess.* Isolated personnel must be continually assessed and monitored throughout the duration of the recovery process. It is important that isolated personnel not deteriorate physically or mentally. It may be beneficial to reassure them occasionally to help maintain morale and focus on their successful return to friendly control.

- *Procedures and contingencies.* The UAR force should thoroughly brief isolated personnel on procedures, restrictions, and recontact plans while isolated personnel are in the custody of the recovery force.

- *Security.* Direct interface between isolated personnel and the UAR force should be strictly limited to preserve OPSEC and the future viability of the UAR recovery force. The logistical support required to sustain a potentially injured individual while waiting to move him, and during actual movement, without violating OPSEC, will be one of the primary planning considerations of the UAR force.

MOVE

3-72. The movement phase of the recovery process consists of all actions taken to transport isolated personnel from a contact point to an exfiltration site (Figure 3-15, page 3-33). The movement may include multiple legs, multiple methods, or multiple elements. The UAR force must thoroughly brief isolated personnel while in transit regarding procedures, restrictions, and recontact plans. To ensure the future viability of the UAR force and the safety of isolated personnel, OPSEC procedures practiced by the UAR force will impact the speed at which isolated personnel are moved from place to place. The logistical austerity of the environment and the UAR force, and the distance over which the isolated personnel must be transported, will also affect the movement process.

3-73. There may be occasion when the recovery force that made the initial contact with the isolated personnel cannot, for operational reasons or other limitations, deliver those isolated personnel safely to friendly control. In such cases, the isolated personnel may be turned over to another NAR asset or conventional recovery force to complete the extraction from hostile territory. The UARCC is responsible for coordinating all handovers and crossovers of isolated personnel recovered by a UAR force.

EXFILTRATE

3-74. Exfiltration is the final movement of isolated personnel from hostile territory to definitive U.S. Government control in a nonhostile environment. Exfiltration (Figure 3-16, page 3-33) will occur by the most secure means available, be it an exfiltration point serviced by a CSARTF, by clandestine aircraft or watercraft, by ground movement crossing an international border, or by passage of friendly lines.

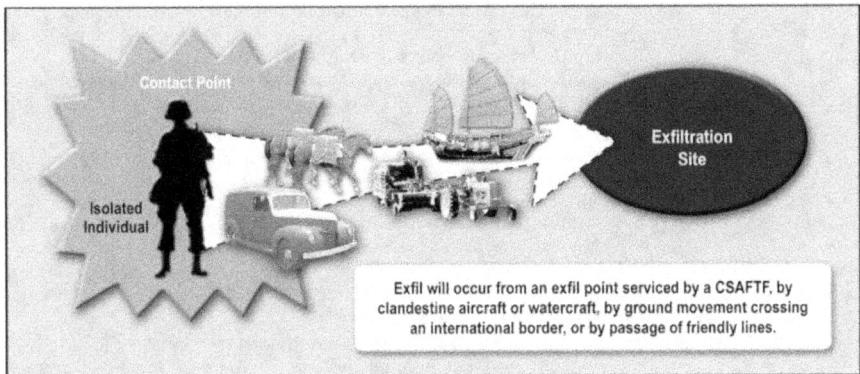

Figure 3-15. Move

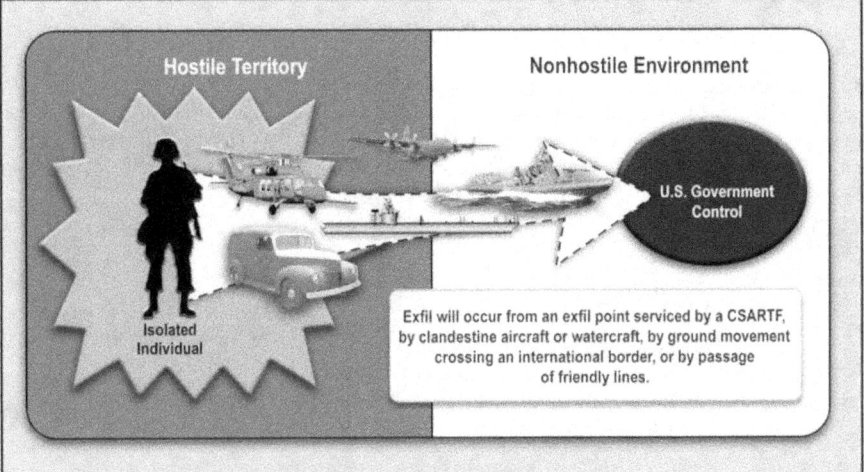

Figure 3-16. Exfiltrate to Friendly Control

3-75. SOF UW ground and maritime forces, other DOD and U.S. Government agencies and allied or coalition forces may provide NAR assets to contact, authenticate, support, move, and exfiltrate evaders to friendly control. UAR forces may be employed to conduct unilateral nonconventional recovery operations or they may work with RMs. These forces develop specific, detailed plans to complement or provide other PR capabilities. SOF UAR forces consist of the following:

- *U.S. Army SF.* UW is the fundamental and foundational mission of SF. In the conduct of UAR, SF units are normally tasked to service coordinated contact points within their assigned area of operation. These units may unilaterally develop or provide support to establish an RM. SF may conduct UAR unilaterally, with indigenous or surrogate assets, or with OGAs.
- *Naval special warfare forces.* Naval special warfare forces are organized and trained to conduct UW operations primarily in maritime, littoral, and riverine environments. Like SF, naval special warfare forces can be infiltrated into their assigned area of operation before or during hostilities to service coordinated contact points. U.S. Navy SEALs may conduct UAR unilaterally, with indigenous or surrogate assets, or with OGAs. SEALs can exfiltrate evaders by submarine, surface vessel, or aircraft.
- *Indigenous or surrogate personnel.* Indigenous or surrogate forces include, but are not limited to, local nationals, guerrilla groups, resistance forces, third-country nationals, or other clandestine organizations. The use of such forces recruited, trained, supported, advised, or led by U.S. SF may provide additional operational flexibility.

RT

3-76. RTs (Figure 3-17, page 3-35) are designated U.S. or U.S.–directed forces that operate unilaterally or with indigenous or surrogate forces and are tasked to contact, authenticate, support, move, and exfiltrate isolated personnel. RTs deploy into uncertain or hostile areas before strike operations in support of the JFC's comprehensive PR plan. RTs may interoperate with other NAR forces and other U.S., allied, or coalition PR capabilities.

UART

3-77. A UART (Figure 3-18, page 3-35) is a designated SOF UW ground or maritime force tasked to contact, authenticate, support, move, and exfiltrate isolated personnel unilaterally or by, with, or through indigenous or surrogate forces. UARTs deploy into uncertain or hostile areas before strike operations in support of the JFC's comprehensive PR plan. UARTs provide the JFC with recovery options in areas where the employment of other PR assets is infeasible or unacceptable or where PR assets are nonexistent. UARTs may interoperate with other NAR forces (OGAs and indigenous or surrogate forces) and with other U.S., allied, or coalition PR capabilities.

- Designated, trained, and directed for PR.
- May be directed to support existing recovery areas in hostile areas.
- May include, but are not limited to, SOF, DOD, OGA, and indigenous or surrogate force options.

Figure 3-17. RT Characteristics

A UART—
- Is specially trained and directed.
- Requires thorough mission analysis, detailed planning, rehearsals, and precise battlespace management.

SOF may conduct UAR—
- Unilaterally.
- With OGAs.
- Through indigenous or surrogate assets.

UART prepares for prestrike employment.

Figure 3-18. UART Characteristics

3-78. A UART is normally an SFODA or a U.S. Navy SEAL platoon that is tasked to perform any UAR mission. The SFODA or U.S. Navy SEAL platoon is trained and equipped to operate in an overt, covert, or clandestine manner in an enemy-held or hostile area for a specified time to contact, authenticate, support, move, and exfiltrate isolated personnel to nonhostile territory. Once a UART has been emplaced, it may remain in its assigned AO for an extended time to support multiple future recoveries. The range of employment options for a UART is limited only by the operational situation. A UART unilaterally conducting a UAR mission must be prepared to interact with, receive from, crossover to, or develop a new recovery mechanism. Therefore, the skill sets are the same as those used by an element working solely with an unconventional assisted recovery mechanism (UARM). Figure 3-19, page 3-36, demonstrates UAR conducted unilaterally by a UART.

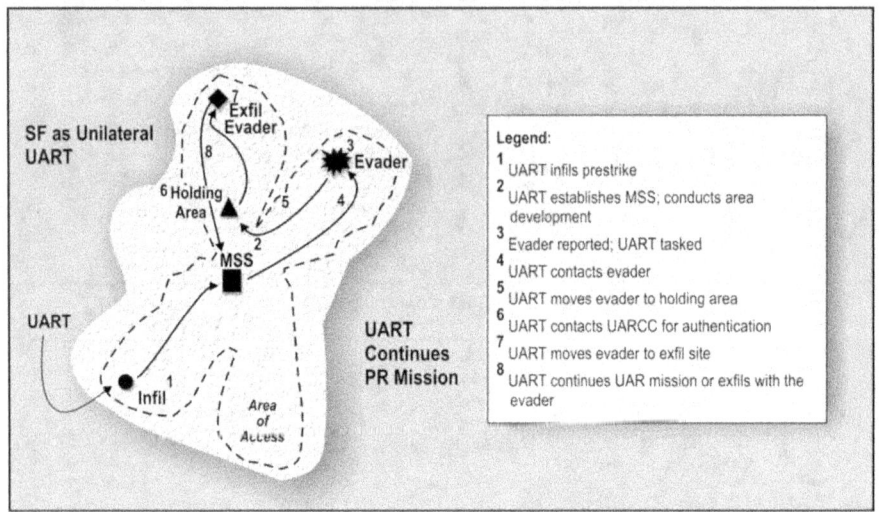

Figure 3-19. Unilateral UART

RM

3-79. An RM (Figure 3-20) is a designated indigenous or surrogate infrastructure specifically developed, trained, and directed to contact, authenticate, support, move, and exfiltrate designated isolated personnel from uncertain or hostile areas to friendly control. RMs may interoperate with other U.S., allied, or coalition PR capabilities. The term "RM" has replaced the obsolete term "escape and evasion nets" and its definition.

- Operate with OGAs and surrogate or indigenous assets.
- Have a designated infrastructure.
- Are trained and directed for PR.
- Are covert or clandestine.
- Include, but are not limited to, indigenous or surrogate assets controlled by the U.S.

Figure 3-20. RM Characteristics

VALUE OF AN RM

3-80. The development and emplacement of an RM takes a great deal of time, money, and effort. RMs are valuable assets, because they can support and supplement conventional and unconventional recovery operations. Their potential value should always be considered and, whenever appropriate, incorporated into recovery planning. With proper planning and support, RMs can be established in almost any environment.

SPECIAL PROCEDURES FOR AN RM

3-81. Successful RM operations require special procedures. RM operations cannot be conducted under the same principles as conventional military operations. When recovery planners plan RM operations, all security, communications, intelligence, diplomatic, command, supply, transportation, fiscal, and legal procedures should be consistent with the goals of returning evaders to friendly control, while ensuring the continued survival of the RM.

ESTABLISHMENT OF AN RM IN ADVANCE OF OPERATIONS

3-82. RMs should be created and maintained in advance of their potential need, or they will most likely not be available to the JFC. The process of establishing viable RMs is long, hazardous, and expensive. No amount of last-minute command interest or sudden infusion of resources can expedite the process without seriously jeopardizing the security of the RM. Constant support of RMs at certain minimal levels is ultimately more successful and less expensive than sudden, sporadic support.

UARM

3-83. RMs include, but are not limited to, UARMs (Figure 3-21). UARMs are designated indigenous or surrogate infrastructure specifically developed, trained, coordinated, and advised by SOF UW ground and maritime forces to contact, authenticate, support, move, and exfiltrate designated isolated personnel in uncertain or hostile areas to friendly control.

- UARM elements are trained and/or controlled by specially trained SOF personnel.
- Unilateral UARM is conducted IAW Title 10, USC, and Director of Central Intelligence Directive (DCID) 5/1.
- SOF-directed UARM interfaces with existing surrogate forces to contact, authenticate, support, move, and exfil.

Figure 3-21. UARM Characteristics

3-84. UARMs may interoperate with other NAR forces and other U.S., allied, or coalition capabilities. A UARM encompasses SOF activities related to the creation, coordination, supervision, C2, and employment of RMs in support of combatant commands. UARMs may include the use of an RT or a UART.

3-85. SF units provide the greatest potential for assistance to theater PR capabilities when acting as a force multiplier and the C2 element for a UARM. These RMs, consisting of indigenous or third-country personnel, perform PR operations in hostile or denied areas by using their inherent placement and access, relative freedom of movement, and plausible deniability derived from their daily activities. SF provides the JFSOCC and JFC with C2 linkage to activate and direct the mechanism during the conduct of UAR operations.

3-86. SFODAs tasked to conduct UAR operations within a designated unconventional warfare operating area (UWOA) or JSOA may be tasked to develop a UARM of indigenous personnel. Figure 3-22 shows a UAR operation conducted by a UARM developed and directed by an SF UART.

3-87. A UART may conduct UAR operations with an OGA. Figure 3-23, page 3-39, shows an OGA RM conducting a crossover operation with a UART.

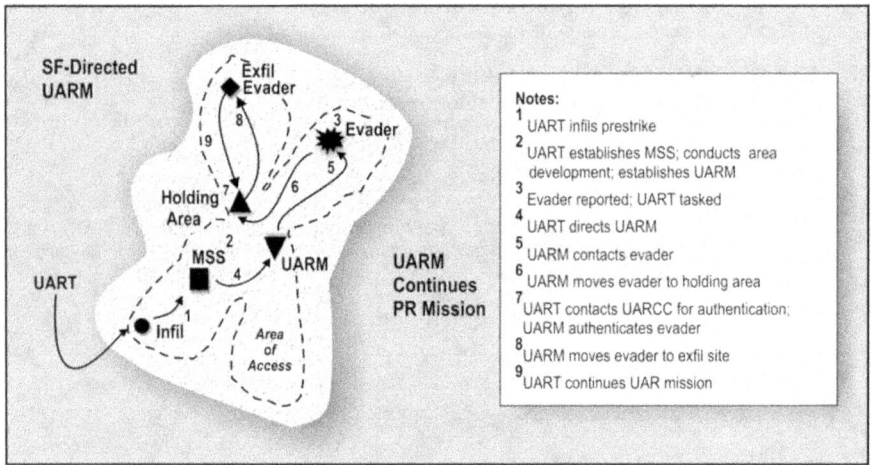

Figure 3-22. SF UART Directs UARM

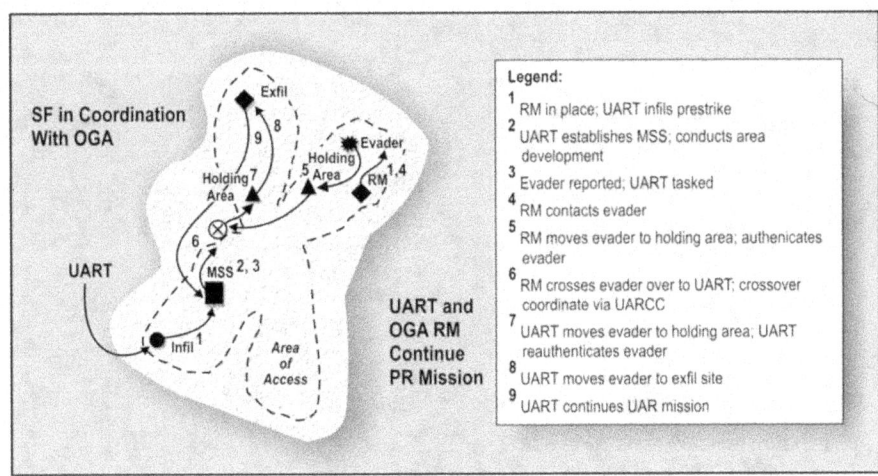

Figure 3-23. OGA RM Crossover to UART

OPERATION MARKET GARDEN

3-88. The following is a vignette of Operation MARKET GARDEN (Figure 3-24, page 3-40), which was conducted during World War II. Operation MARKET GARDEN highlighted the important role OSS Jedburgh teams (Figure 3-25, page 3-41) played in rescue efforts during this operation.

In 1944, the Allied advance on continental Europe was designed to ultimately cut off the retreating German forces in north Holland and then swing eastward to take the Ruhr Valley to cut off the German war machine from the German industrial heartland. Eisenhower reluctantly agreed with Montgomery that an airborne invasion would deal a final, paralyzing blow to the Germans on the western front. Operation MARKET GARDEN began at midnight 16 September 1944 and was projected to last 5 days.

The British 1st Airborne Division would target the village of Arnhem. The American 82d Airborne Division would target Nijmegen, and the American 101st Airborne Division would target Eindhoven. The tanks of the British XXX Corps, leading the 2d British Army, would join up with the Paras.

What is not so well known is the role played by OSS Jedburgh teams in support of Allied operations in the region. Two different Jedburgh teams (Figure 3-25, page 3-41) significantly contributed to the recovery of more than 200 evading paratroopers for months following the airborne invasion around Arnhem.

In June 1943, a Jedburgh team led by Dick Kragt, was infiltrated into the area. His mission was to provide support to evading Allied airmen conducting operations on the European continent. Kragt made contact with the Dutch Resistance and operated from Ede.

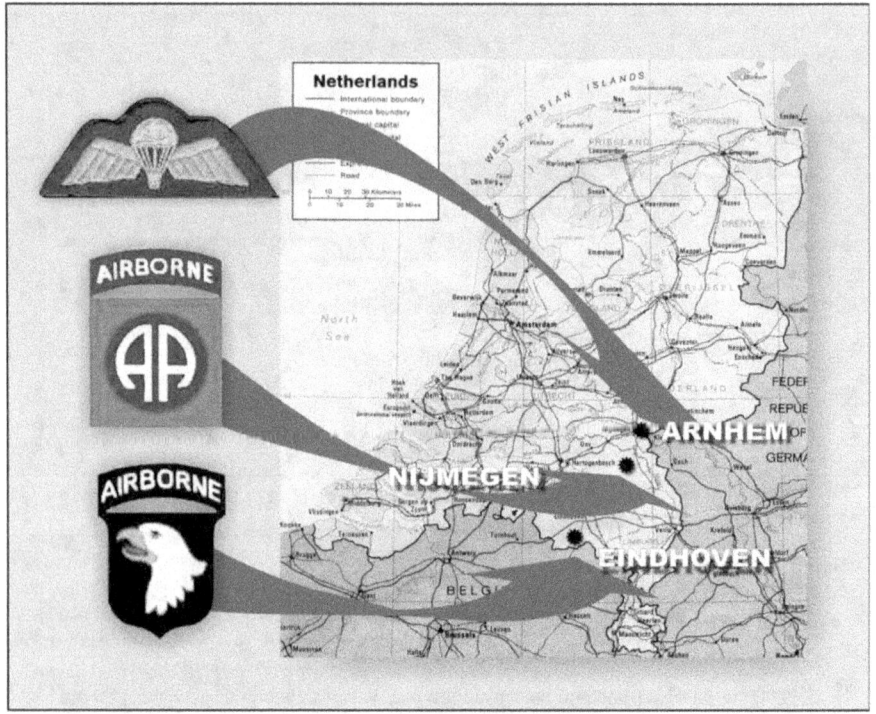

Figure 3-24. Operation MARKET GARDEN

During infiltration into his operational area, he barely missed a church steeple, brushed against a large poplar tree, and tumbled into a ploughed turnip field. All his equipment, including his radio set, landed miles away and was retrieved by the Germans. Kragt, without his radio set, with little money, and with a .45 pistol, did, however, successfully link up with Joop Pillar of the Dutch Resistance. With little more than the clothes on his back, but with the aid of the Dutch Resistance, Kragt was able to complete his complex mission of infrastructure organization to support evaders. Beginning in September 1944, Kragt would interoperate with another Jedburgh team. In early September 1944, another Jedburgh team was infiltrated into the area. The team consisted of Gilbert Sadi-Kirschen, Jules Regner, Rene Pietquin, and Jean Moyse. Their primary mission was intelligence collection, but with the initiation of MARKET GARDEN, the team was pressed into support for Allied evaders.

Piet Een and his efficient reception committee, who moved the team to a local hunting lodge, met the Jedburgh team on their DZ. The next day, Kirschen was introduced to Jan van der Ley, the Resistance chief. Roelof van Valkenburg, who would act as the liaison between the Jedburgh team and the Resistance, and Bep Labouchere, who would act as their courier, accompanied Jan.

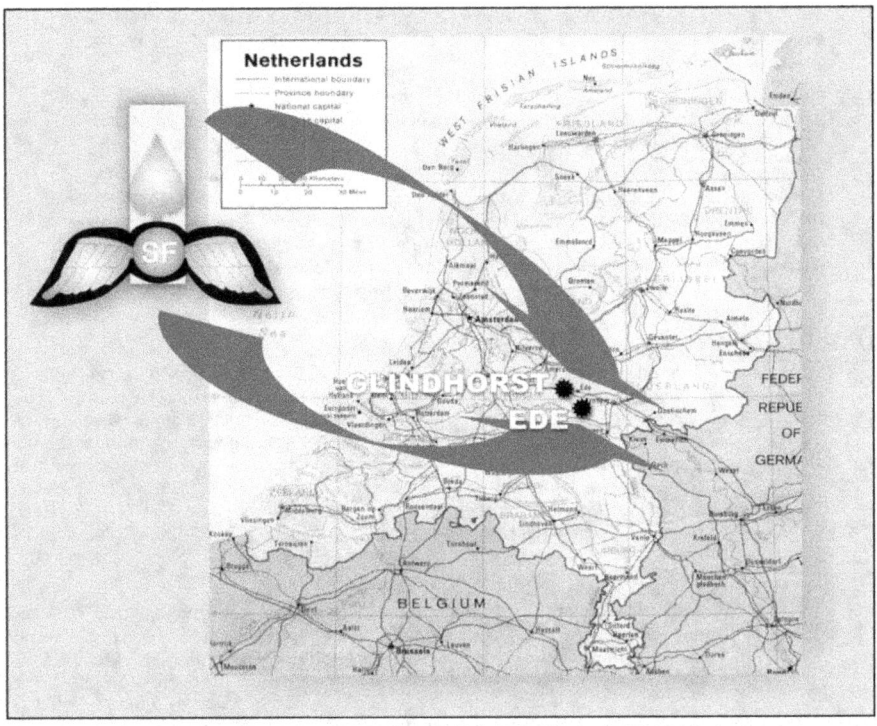

Figure 3-25. Operation MARKET GARDEN, the Jedburghs

Kirschen took up residence in a chicken coop at the farm of Wuf Langerwey outside the village of Glindhorst. Kirschen and Van Valkenburg established their intelligence collection system. Messages passed throughout the system daily.

On 16 September, Kirschen received official notification of the beginning of Operation MARKET GARDEN via radio traffic. Radio traffic between the 1st Airborne Division and the War Office in England was suddenly lost, and in the early hours of Wednesday morning, Kirschen, reacting to instructions, dispatched an element to reestablish contact with the airborne troops at Oosterbeek.

During Operation MARKET GARDEN, and continuing into its aftermath, Edith Nijhoff acted as a transportation asset using her horse cart. Edith Nijhoff roamed the countryside by day, and frequently could be seen driving a horse and cart piled high with sacks and old blankets. Underneath the sacks were evaders being transported from one hiding place to another.

Jimmy Edwards and three members of his Dakota crew of the 271st Royal Air Force (RAF) Squadron were shot down near Nijmegen. The Dutch Resistance hid these survivors in the surrounding woods. A local doctor treated their severe burns. They were then transported to the house of a local priest, where they stayed until the advancing Allies overran their location.

George Sykes, a P-51 pilot of the 479th Fighter Group USAF, was shot down near Arnhem. Bill Wlideboer, of Ede, and members of the Resistance recovered Sykes and hid him in a mental hospital in Wilheze.

Continuing from October through December, Kirschen, his Jedburgh team, and the Dutch Resistance continued to support the recovery of Allied forces pursuant to Operation MARKET GARDEN. They provided shelter, clothing, medical treatment, food, and transportation and returned hundreds of evaders to friendly control. They continued to support captured Allied forces who had escaped from captivity in German detention camps in the surrounding area and were making their way back to friendly control.

UAR PLANNING

3-89. Deliberate planning for UAR operations begins with the geographical combatant commander's vision of prioritized areas in which he and his staff expect the likelihood of U.S. involvement. Development of these plans generally follows an established procedure beginning with the Joint Strategic Capabilities Plan (JSCP). The JSCP provides guidance to the combatant commanders and the service chiefs to accomplish tasks and missions according to their current military capabilities. The JSCP apportions resources to combatant commanders according to military capabilities resulting from program and budget actions. The JSCP provides the framework for advice provided to the Secretary of Defense.

3-90. UAR planning is derived from information outlined within OPLAN and component supporting plans. The SOC commander receives mission taskings and, with his staff, outlines requirements that specifically task apportioned SOF forces. These taskings are outlined in the mission letters for the JSCP. USSOCOM receives the SOF mission letters and must review and validate them before supplemental mission letters are produced for subordinate SOF units. The evolution process, which produces mission letters defining SFODA taskings, is derived from annual targeting conferences, the master attack plan, NAR requirements planning, OPLAN reviews, and CSAR efforts to identify PR requirements and provide PR planners with the basis for identifying potential gaps in the overall PR plan of the theater. SOF commanders from apportioned and allocated forces will use all available information obtained in previous planning and conduct campaign planning.

3-91. Using the results from all planning efforts, the next logical step is to conduct NAR planning. NAR planners will identify areas within the JSOA or supporting areas where conventional PR assets are unavailable or infeasible or where complementary coverage is required, because limited asset availability is expected. This information will direct the SOF planners toward the NAR requirements that should be developed to best support the overall PR plan. This plan will incorporate military and OGA capabilities.

3-92. The combined efforts of all theater planning coupled with the SOF campaign planning results will determine what UAR taskings will be included on the supplemental mission letters produced by the SF group commander and his staff. Subordinate commanders will receive mission letters derived from this planning. The mission letters will articulate the missions, tasks, and concepts that the SF group must be capable of executing in support of the specific plan. SF group planners will provide their forces with supplemental mission letters outlining the responsibilities of the SFODA.

3-93. When tasked to conduct deliberate planning for UAR missions, SFODAs must conduct detailed mission analysis of the assigned operational area. The SFODA should receive current detailed intelligence. The SFODA should receive all available detailed information on identified recovery support areas and the lack of conventional PR capability specifically outlined within comprehensive theater PR plans. The SFODA should identify, according to the threat, those assigned areas where unilateral operations are feasible within the JSOA. SFODAs tasked to conduct UAR operations within a designated UWOA or JSOA may be tasked to develop a UARM that can include indigenous personnel. Development of UARMs will be conducted IAW Title 10, USC; DCID 5/1; and other appropriate regulations.

3-94. To address any area deemed unsuitable for unilateral operations, the SFODA should identify the requirement for infrastructure development through their planning staffs and chain of command. Requirements should specifically articulate critical tasks to be performed, necessary areas of access, number of evaders to be accommodated simultaneously, and duration of support. The SFODA does not have to wait for the completion of formal theater planning before identifying requirements that will affect their mission accomplishment. In many cases, shortfalls identified at the detachment level will have an affect on the decision made at the JFC planning level.

3-95. This process is much simpler when the SFODA has mission tasking outlined within their mission letter, information obtained from theater targeting and NAR planning, and a clear understanding of the potential role they will play in the overall PR plan. The SFODA's planning efforts can be used to identify entities within the overall PR infrastructure. The SF group's representative at the theater NAR planning conference will present this information.

3-96. When a requirement is forwarded for action, the JFSOCC, as the JFC's OPR for NAR, crafts a requirements message for the geographic combatant commander. The message is then coordinated with the combatant commander's J-2, J-3, and PR OPR. Once the geographic combatant commander approves the requirements message, the message is transmitted to the joint staff for validation and action. The joint staff PR OPR passes the requirement through appropriate channels to the intelligence community for a feasibility assessment and CONOPS development.

UAR ASSETS

3-97. UAR assets are as follows:

- *Guerrillas.* The use of guerrillas to recover evaders may provide added operational flexibility. Depending on the size of the guerrilla groups and their territorial control, they may be free to operate more overtly and can control or limit enemy activity in the group's operational area. For these reasons, evaders do not pose as great a security threat to guerrilla groups as they do to SF teams. In addition, there may be less of a requirement to quickly exfiltrate evaders from these groups. There are two types of guerrilla groups—

- *Sponsored.* U.S. or allied SF elements may support, lead, or advise sponsored guerrilla groups. Friendly forces may recruit and train guerrilla groups, or guerrilla groups may be dependent on allied countries. Because exfiltration of personnel and materiel may be a routine operation for these groups, evaders may be more expeditiously returned to friendly territory through their assistance. Planners should ensure guerrilla groups are able to correctly use the communications, contact, and authentication procedures that have been established for the theater of operations. Evaders who find themselves under the control of such guerrilla groups should comply with all reasonable instructions issued by the group.

- *Unsponsored.* Unsponsored, independent guerrilla groups may be comprised of mercenaries, dissidents, and outlaws. They may assist evaders when it is in the group's perceived best interest, for example, through the convincing lure of a blood chit reward. PSYOP may be employed to convince such groups it is in their best interest to aid evaders. (Chapter 5 provides more information on the role of PSYOP in PR.) Because it is unlikely that independent groups will have been trained in the same communications, contact, and authentication procedures as sponsored groups, planners trying to make use of such groups should be prepared to employ modified or alternate procedures. The use of unsponsored guerrilla groups to support recovery operations presents certain problems. Communications limitations between friendly forces and these guerrilla groups can increase the difficulty in arranging the recovery of evaders, thereby extending the evasion period. Under certain conditions, the group may try to use the evader to augment its forces. This situation presents problems, because guerrilla groups may conduct operations in violation of U.S. policy or international law. Under these conditions, evaders are to resist guerrilla attempts to solicit their participation and, if forced to participate, avoid direct involvement or minimize the effects of such actions. Additionally, such groups may perceive advantages in retaining an evader for extended periods for use as a bargaining chip. Because independent guerrilla groups may perceive the evader as a de facto representative of the U.S. Government, evaders should conduct themselves with the utmost discretion.

- *Clandestine organizations.* Clandestine organizations are comprised mainly of indigenous personnel operating clandestinely and/or covertly in the hostile territory. Clandestine organizations are engaged in activities designed to change the political or military situation in that territory. These organizations may be comprised of political dissidents or minority groups that support resistance, revolution, or friendly intelligence activities by collecting, hiding, and forwarding materiel, information, and personnel.

- *Specially equipped aircraft.* When an evader's exact location is known and is within flying range of helicopters and/or fixed-wing aircraft, a single, specially equipped aircraft can carry out a recovery. This aircraft should avoid enemy detection, quickly enter the enemy-controlled area, recover the evader, and return. Helicopters with an air-refueling capability, night vision devices (NVDs), and terrain-following radars can carry out a recovery.

Chapter 4

Special Forces Liberation

The DOD, according to its stated policy, has a moral obligation to protect its personnel, prevent exploitation of its personnel by its adversaries, and reduce the potential for captured personnel to be used as leverage against the United States. The recent detention incidents of the 1990s were big stories because of the detained Americans: Chief Warrant Officer 3 (CW3) Hall, a U.S. pilot shot down in North Korea; CW3 Durrant, a 160th SOAR helicopter pilot shot down and detained in Mogadishu; and, most recently, the three U.S. Army PWs detained in Kosovo. These incidents, when handled expeditiously, can enhance military operations as a whole. However, when these incidents are poorly handled, they can degrade political and civil support for operations. The enemy can be denied a source of intelligence by liberating detained personnel in an expeditious manner to preclude enemy exploitation. The mere fact that American Service members know the U.S. Government will exhaust all efforts to liberate them from captivity is a big morale booster and could encourage detained personnel to resist enemy exploitation.

DA OPERATIONS

4-1. DA operations are normally limited in scope, requiring an SFODA to infiltrate a denied area, attack a target, and conduct a preplanned exfiltration. The preplanned exfiltration may include long-term, stay-behind operations. DA operations achieve specific, well-defined, time-sensitive results of strategic or operational significance. They normally occur beyond the range and operational capabilities of tactical weapons systems and conventional maneuver forces. These operations may require the knowledge of often regionally oriented SF-unique skill sets developed in support of UW. SF DA operations may be unilateral or multinational, but they always occur under a U.S. chain of command. SF units can conduct these operations across the full spectrum of conflict at the operational or strategic level in support of the JFC and component commanders. In DA operations, SF units may be tasked to conduct PR operations.

LIBERATION OPERATIONS

4-2. When directed by the President or the appropriate combatant commander, SF units may be tasked to conduct liberation operations to recover individuals from enemy control. SF units are specifically organized, trained, and equipped to conduct DA missions and would employ DA TTP to conduct liberation operations. SF liberation operations are DA missions, not PR missions. A DA mission is normally planned as a one-time surgical operation against a single target. Usually planned and rehearsed outside the area of responsibility (AOR) under the direct supervision of the SECDEF and the President. DA missions are usually conducted by a specific JSOTF. Thus, they occur outside the PR C2 structure.

4-3. Liberation operations elements must prepare an EPA and ensure it is developed and coordinated with the appropriate JFC's PR structure. In addition, those personnel rescued or recovered from captivity or detention will require processing through the established repatriation plan.

4-4. The Son Tay raid during the Vietnam War is an example of SF used in a liberation operation. The following vignette discusses the mission, and Figure 4-1, page 4-3, shows the Son Tay Camp.

By early 1970, there were more than 450 known American PWs in North Vietnam and another 970 American soldiers who were MIA. Some of the PWs had been held in captivity for more than 5 years, enduring ghastly conditions, rough treatment, torture, and torment. Some PWs even died. In May 1970, reconnaissance photographs showed ground-to-air signals emplaced by U.S. PWs, revealing the existence of two prison camps west of Hanoi. Further reconnaissance photographs taken by SR-71 Blackbirds revealed that Son Tay was an active prison camp. The camp, in the open and surrounded by rice paddies, was near the 12th North Vietnamese Army (NVA) regiment totaling about 12,000 troops. Additionally, Phuc Yen Air Base was only 20 miles (32.2 km) northeast of Son Tay, and it was determined that the camp was being enlarged because of increased activity. Any rescue effort would have to be executed swiftly. If not, the enemy could have planes in the air and a reactionary force at the camp within minutes.

Brigadier General Donald D. Blackburn suggested a small group of SF volunteers rescue the PWs. He chose Colonel Arthur D. "Bull" Simons to lead the group. At Fort Bragg, North Carolina, Colonel Simons chose 100 dedicated SF volunteers that possessed the requisite skills to pull off the rescue mission. These volunteers were trained in isolation, using a full-scale replica of the Son Tay Camp constructed at Eglin Air Force Base, Florida, to conduct critical rehearsals for the mission. In November 1970, the Son Tay raiders moved to Takhli, Thailand, a CIA-operated secure compound for final preparation. Of the original 100 SF members trained, 56 were selected for the mission. Only Simons and three others knew the true mission. Five hours before takeoff on November 10, Simons told his 56 men: "We are going to rescue 70 American prisoners of war, maybe more, from a camp called Son Tay. This is something American prisoners have a right to expect from their fellow soldiers. The target is 23 miles (37 km) west of Hanoi."

The plan was not unduly complicated. Using in-flight refueling, the six helicopters would fly from Thailand, across Laos, and into North Vietnam. While various diversions were taking place locally and across North Vietnam, the task force would close on the camp under cover of darkness. A single HH-3H with a small assault element would be crash-landed inside the prison compound, while two HH-53s would disgorge the bulk of the assault force outside. The wall of the camp would be breached and the prison building stormed. Any NVA troops found inside would be killed and the PWs would be taken outside and exfiltrated via the helicopters. On 21 November, at about 2318, the Son Tay raiders commenced the rescue mission. Except for some minor difficulties, the raid went according to plan. However, the raiders soon realized that there were no PWs at the camp. It was later learned that the PWs had been relocated to Dong Hoi on 4 July. The PWs were moved by their captors because of a flooding problem at the camp, not because they had learned of a pending rescue attempt.

Despite the intelligence failure, the raid was a tactical success. The assault force got to the camp and took their objective. Although no PWs were rescued, the raid sent a clear message to North Vietnam that Americans were outraged at the treatment U.S. PWs were receiving and would go to any length to bring their men home. Aware of the raid on Son Tay, the morale soared for the moved prisoners for they now knew that America cared and attempts were being made to free them. The raid triggered subtle, but important, changes in the treatment of American PWs by the Vietnamese.

Figure 4-1. Son Tay Camp

SF DA MISSION IN SUPPORT OF PR

4-5. SF DA recovery operations are short-duration surgical strikes and other small-scale offense actions conducted to locate, recover, and return to friendly control personnel or material held captive, isolated, or threatened in areas sensitive, denied, or contested. Detailed planning, rehearsals, and thorough intelligence analysis often characterize these SF recovery missions. These operations typically leave a smaller signature than conventional operations and can create effects disproportionate to the size of the committed force. The need for precision, combined with requirements for other SF-unique capabilities, often makes SF the choice for a myriad of DA tasks, including support to PR. The small size and limited firepower of SFODAs makes DA recovery operations rely on the synergistic effect of speed, surprise, violence of action and, oftentimes, the cover of darkness.

Chapter 5

Civil Affairs and Psychological Operations Support to Personnel Recovery

CA soldiers have an inherent role in early PR planning and intelligence analysis. During the SOF mission planning process and the subsequent development of a PR contingency plan, CA soldiers are an essential information source. Specifically, CA input to the new JPRSPs can aid SOC planners with pragmatic, timely data.

The mission of PSYOP is to plan and conduct PSYOP to support military operations, to support the attainment of U.S. national objectives abroad, and to be prepared to conduct such operations unilaterally and/or with other military services and U.S. Government agencies. Psychological preparation of the PR and/or UAR AO is essential to set the conditions for RMs.

CA SUPPORT

5-1. CA units are a good source of information. As such, they can provide essential information to recovery planners.

TOOLS

5-2. CA units maintain valuable tools to assist in UAR planning. CA subject matter experts may have been deployed to the target area and can elaborate on infrastructure details. Cultural studies highlight the prevailing moods, attitudes, and historical trends of a population. CA units maintain a wealth of information from lines of communication to population needs in their country studies, country surveys, and AARs. Population density studies can, for example, help SOF planners develop an RM feasibility study. After deployment, CA units produce reports that give factual, timely information to the SOF community via the special operations debrief and retrieval system (SODARS). CA-authored SODARS reports consist of the following relevant information:

- Details on critical facilities.
- Political parties and factions.
- Social and economic factors.
- Military.
- Paramilitary.
- Police.
- Demographics.

SYNCHRONIZING

5-3. Once SF elements initiate their EPA or begin a SOF recovery operation, CA units and soldiers can help to synchronize the recovery plan. CA soldiers have direct access to NGOs, private venture organizations, and/or other various governmental organizations that may influence the indigenous environment. CA soldiers augment interagency working groups with their ability to coordinate and focus otherwise diverse organizations toward the common PR mission. Should negotiations become necessary, CA teams may provide negotiators with key information through analysis of the situation and the operational environment. As planners identify possible intermediate staging bases (ISBs), CA teams can bring all players together in the civil-military operations center (CMOC) to centralize resources and provide the greatest unity of effort toward the PR mission.

5-4. CA planners collaborate with the DOS to develop, review, and recommend actions or initiatives to support current Embassy drawdown plans. These plans prioritize personnel for extraction and identify assembly areas; however, these plans do not address UAR requirements beyond the scope of a NEO. Currently, DOD and the DOS are working on a memorandum of agreement (MOA) for mutual support of PR. This MOA will allow the DOS to identify UAR requirements and capabilities needed to recover isolated personnel who cannot make it to the extraction site. UAR requirements could include SF recovery teams assisting in the recovery of American citizens, an NGO, or a private venture organization during NEO. CA planners' knowledge of DOS UAR requirements can ensure a NEO adequately addresses PR contingency issues. Therefore, all PR planning agencies should consider CA knowledge on current NEO plans to capitalize on a prudent PR design.

5-5. CA soldiers and units assist PR planners by using factual, timely, and culturally relevant information. CA soldiers have ongoing relationships with NGOs, private venture organizations, and OGAs that may give negotiators an increased advantage should negotiations become necessary. Finally, CA teams synchronize otherwise diverse players in a PR effort by using the CMOC to augment existing C2 nodes from CONUS to the ISB. Recovery planners can best achieve unity of effort by using CA soldiers and units early in the PR planning process. Finally, the CA component of a solid PR strategy gives planners increased flexibility should operational and tactical situations change.

PSYOP SUPPORT

5-6. PSYOP contribute the following competencies to support PR and/or UAR with specific PSYOP themes and messages designed to—

- Modify the attitude and influence the behavior of selected foreign groups to support the attainment of U.S. national objectives (in the case of PR, to protect and recover U.S. personnel).
- Interrelate closely with national economic, political, and military activities to establish, sustain, and reinforce the effect and credibility of these integrated functions.

- Apply concerted, systematic, and comprehensive PSYOP programs that support the capabilities of U.S. Government agencies and departments concerned with national security affairs.

5-7. PSYOP forces can provide the lead agency or commander with a lean operational asset capable of using "reach-back" techniques for effective product development. These same assets also support military operations in populated areas (for example, man-packed, vehicle-mounted, and helicopter-mounted loudspeakers). In addition, PSYOP forces provide continuity for up-to-date situational awareness across the whole spectrum, from the strategic level to the tactical level.

CONSIDERATIONS

5-8. There are several requirements to consider when using PSYOP forces. PSYOP assets require early integration in the planning process. The PSYOP planner(s) must access operational decision maker(s) to effectively coordinate and orchestrate the PSYOP role. PSYOP often require external logistical support, such as production assets or media support and aviation assets. The supported headquarters must support these requirements.

5-9. Specific areas in which PSYOP forces can assist include the following:

- *Survival, evasion, resistance, escape, recovery (SERER).* The psychological objectives in this situation are to convince people to provide assistance to escaping and evading personnel. PSYOP specialists base the appeals on specific conditions and attitudes that can influence the indigenous noncombatants in the AOR. Alternatively, they can warn of retribution if evading personnel are mistreated or captured.
- *Search and rescue (SAR) and CSAR.* PSYOP support SAR and CSAR as communicators to the indigenous population, provide translator support, assist in crowd control, and provide cultural guidance (do's and don'ts).
- *Forced recovery.* PSYOP forces offer nonlethal fire to help reduce interference by noncombatants and to minimize resistance by combatants.
- *Deception.* PSYOP can help to facilitate concealment of the timing and method of CSAR or UAR operations with deception planners.
- *Negotiation.* PSYOP forces are typically accustomed to working at the interagency level under the direction of the United States Information Agency, DOS, or OGAs and with NGOs. PSYOP can be an integral part of public diplomacy and, when required, prepare and coordinate multidisciplined public diplomacy and information campaigns.
- *Postmission information operations.* PSYOP forces can publicize the blood chits program and encourage its continued support. They can support personnel without blood chits by encouraging cooperation (positive publicity) and by generating sympathy and support among the populace in the AOR. PSYOP personnel will highlight successes and downplay the failures of missions and operations. An effective blood chit program in the theater, coupled with an active PSYOP effort, may

encourage these types of unconventional recoveries and, if properly exploited, could eventually become an organized RM.
- *General support.* PSYOP provide special psychological operations assessments (SPAs) that offer critical PSYOP intelligence on how the indigenous populace will deal with isolated personnel either while in captivity or in the process of escaping. PSYOP forces can develop pointee-talkees for use by isolated personnel for basic communications with the indigenous population.

SUPPORT

5-10. A planned PSYOP program is essential in achieving cohesive efforts between the SOF conducting the UAR and the indigenous assets supporting PR. The appeals designed to obtain this kind of support may be developed around the following ideas:

- Humanitarian acts.
- Rewards or monetary benefits (supports theater blood chit program).
- Allied unity or patriotism.
- Retaliation against the enemy.

CAPABILITIES

5-11. PSYOP can provide an analysis of enemy propaganda and psychological exploitation techniques used against individuals in captivity to develop a resistance posture. Propaganda analysis is the detailed examination of the source, content, audience, media, and effects (SCAME) of a propaganda message to obtain intelligence that supplements the conventional forms of analysis. Some SERE contingency guides identify propaganda exploitation techniques used by the targeted country. PSYOP can also develop a counterpropaganda program to support individuals in captivity, thus easing the effects of psychological and propaganda exploitation.

5-12. PSYOP can also use other assets to support a theater PSYOP campaign. The AFSOC EC-130 (known as COMMANDO SOLO) is such an asset. COMMANDO SOLO is an airborne electronic broadcasting system using four EC-130E Rivet Rider aircraft operated by the 193d Special Operations Group, Pennsylvania Air National Guard. COMMANDO SOLO conducts PSYOP and CA broadcast missions in the standard amplitude modulation (AM), frequency modulation (FM), high frequency (HF), television (TV), and military communications bands. The 193d Special Operations Group flies missions at maximum altitudes possible to ensure optimum propagation patterns. Other uses for COMMANDO SOLO include—

- Mask radio frequencies by directing radio frequency energy toward the adversary's receivers, preventing interception of friendly transmissions.
- Broadcast disinformation and deception; for example, playing tapes of a false CSAR operation, to make enemy forces believe PR forces are looking for isolated personnel in a different location.

- Broadcast the value of U.S. personnel and obtain the support of indigenous noncombatants in the AOR.
- Jam or disrupt enemy forces, interfere with the communications of the enemy search forces, and disrupt enemy C2.
- Provide high-power transmission for the C2 of PR assets. The JSRC must provide a PR specialist for this mission, as COMMANDO SOLO aircrews are not trained to understand all the nuances of PR C2.
- Improve radio coverage for communication with isolated personnel. The EC-130 flies during day or night scenarios and can be refueled in flight. A typical mission consists of a single-ship orbit that is offset from the desired target audience. The targets may be either military or civilian personnel. Secondary missions include command, control, and communications countermeasures and limited intelligence-gathering.

Appendix A
Evasion Plan of Action Format

The EPA of an SFODA is of prime importance to the overall successful accomplishment of its mission. Throughout the history of SF, plans for E&E have been a part of every mission. Many great examples of success and well-designed formats exist, but standardization has never occurred. EPA formats, procedures, and SOPs vary considerably from theater to theater, SF group to SF group, and SFODA to SFODA. The following format, excluding samples of the listed appendixes, provides all with a uniform standard for EPA planning.

The intent of this appendix is not to instruct how to write an EPA, but to provide a proven format to assist the SFODA during mission planning. The overall objective is to institute a standard EPA planning and execution process that will ensure continuity across the spectrum of SO.

NOTE: Bold text in the EPA format example identifies those portions of the EPA format that are unit- or mission-specific.

FM 3-05.231

(_____)
CLASSIFICATION

ANNEX W (EVASION AND RECOVERY) to OPERATIONS ORDER or PLAN (Number or Code name), JSOA XXXXX, Co A, 2d Bn, 9th SFG(A)

References: **Map sheet scale, major geographic area, sheet name and number, and series and edition; manuals; ARs; and other documents used to prepare the order or plan must be available to the EPA planner or supplied with the order. Order or plan of next higher HQ is NOT a reference. A SAID is a reference. Show references in block type.**

Time zone used throughout the order or plan: **Spelled out phonetically, in capital letters (for example, ZULU).**

Task organization: **Self-explanatory.**

1. ( ) SITUATION:

   a. ( ) Enemy Forces: **Expand on enemy situation with emphasis on enemy activity throughout evasion boundaries, to include population control measures, police and militia capabilities, and demographics related to attitudes of civilians toward U.S. military. Include probable COAs of all members of the population (for example, military, police, and civilian).**

   b. ( ) Friendly Forces:

   (1) ( ) Location, mission, and planned actions of units that evaders may come in contact with: **Include locations of all civilian and governmental agencies, friendly or neutral embassies, liaisons, or consulates.**

   (2) ( ) Fire support: Cover restrictive fire areas (RFAs) and TRPs along the corridor **(for example, all scheduled stops will be designated RFAs).**

   (3) ( ) Attachments and detachments: **Include aircraft type, call signs, mission number, unit ID, crew manifest, PLS codes, blood chit numbers, and so on for infiltration and exfiltration missions, if known. Include escort aircraft, too.**

   (4) ( ) Assumptions: **List pertinent assumptions made by the SFODA.**

2. ( ) MISSION: (Who, what, when, where, and why.)

"SFODA ___ will conduct long-range/short-range, assisted/unassisted evasion from (vicinity) _____ to _____ (list specific location/grid, and so on) in order to (link up with friendly forces, conduct extraction, link up with UART/UARM, and so on) at (NLT, NET, DTG, and so on)."

3. ( ) EXECUTION:

   a. ( ) Concept of the Operation: The overall plan. (Refer to Appendix 1, EPA Overlay, and Appendix 3, Contact Procedures and Linkup.) **Provide brief statement of how the SFODA will accomplish evasion from beginning to end.**

   (1) ( ) Scheme of maneuver: **Cover locations, descriptions, time windows, and actions at all sites.**

   (2) ( ) Criteria for activation of EPA: **List specific criteria that will cause SFODA to initiate evasion. Examples: On order from SFOB, FOB, AOB, or MSS; attack on JCO**

(_____)
CLASSIFICATION

(_____)
CLASSIFICATION
ANNEX W (EVASION AND RECOVERY) to OPERATIONS ORDER or PLAN (Number or Code Name), JSOA XXXXX, Co A, 2d Bn, 9th SFG(A)

house; loss of rapport with indigenous forces; detachment initiated; missed radio contacts; mission compromise; SFODA is no longer mission-capable; missed exfiltration; and so on. Be sure to include any criteria provided in the FOB E&R Guidance.

(3) ( ) Actions during ingress and egress: (Refer to Appendix 1, EPA Overlay.) **Cover activation of EPA if aircraft is forced to land during either infiltration or exfiltration. Address short- and long-term evasion scenarios, before and after the MDL, and in friendly and enemy territory.**

(a) ( ) Ingress evasion plan:

1) ( ) Flight route: **List locations of departure airfield, spider routes, en route checkpoints, MDL, initial penetration point of enemy airspace, and infiltration LZ or DZ.**

2) ( ) Actions for downed aircraft: **Use subparagraphs to address each leg of infiltration (over water legs, on water legs, and/or over land legs).**

3) ( ) Evasion plan before the MDL:

a) ( ) Joint evasion goals: **Cover immediate and extended evasion intentions.**

b) ( ) Evasion movement procedures: **Address movement with and without infiltration crew.**

c) ( ) Rendezvous (RV) point or hole-up (HU) site: **Explain hiding intentions.**

d) ( ) Actions at RV or HU site: **Explain intended actions, and determine length of stay at the sites.**

e) ( ) Travel intentions: **Address travel distances, duration of movements, time of day for movement, and so on.**

4) ( ) Evasion plan beyond the MDL.

a) ( ) Joint evasion goals: **Cover immediate and extended evasion intentions.**

b) ( ) Evasion movement procedures: **Address movement with and without infiltration crew.**

c) ( ) RV point or HU area: **Explain hiding intentions.**

d) ( ) Actions at RV point or HU site: **Explain intended actions, and determine length of stay at the sites.**

e) ( ) Travel intentions: **Address travel distances, duration of movements, time of day for movement, and so on.**

(b) ( ) Egress flight route: **List locations of mission exfiltration PZ, en route checkpoints, EDP(s), and destination. An EDP should be determined as any point where the egress platform exits the SFODA evasion corridor or boundaries of the SFODA operating area.**

1) ( ) Evasion actions before EDP: **Consider the SFODA EPA versus the evasion plan of the egress platform.**

(_____)
CLASSIFICATION

(_____)
CLASSIFICATION

ANNEX W (EVASION AND RECOVERY) to OPERATIONS ORDER or PLAN (Number or Code name), JSOA XXXXX, Co A, 2d Bn, 9th SFG(A)

    a) ( ) Joint evasion goals: **Cover immediate and extended evasion intentions.**

    b) ( ) Evasion movement procedures: **Address movement with and without exfiltration crew.**

    c) ( ) RV point or HU site: **Explain hiding intentions.**

    d) ( ) Actions at RV point or HU site: **Explain intended actions and determine length of stay at the sites.**

    e) ( ) Travel intentions: **Address travel distances, duration of movements, time of day for movement, and so on.**

    2) ( ) Evasion actions after EDP: **SFODA should normally rely on the evasion plan of the recovery force. Tactical control, once on the ground, must be addressed.**

    a) ( ) Joint evasion goals: **Cover immediate and extended evasion intentions.**

    b) ( ) Evasion movement procedures: **Address movement with and without exfiltration crew.**

    c) ( ) RV point or HU site: **Explain hiding intentions.**

    d) ( ) Actions at RV point or HU site: **Explain intended actions and determine length of stay at the sites.**

    e) ( ) Travel intentions: **Address travel distances, duration of movements, time of day for movement, and so on.**

    (c) ( ) Evasion actions in a permissive environment: **List specific actions to be taken should evasion be necessary in a permissive area. Be sure to include the technical communications plan.**

    (d) ( ) Evasion actions in a nonpermissive environment: **List specific actions to be taken during evasion in hostile areas. Be sure to include the technical communications plan.**

    (4) ( ) Initial evasion goals: **(List actions within first 48 hours of evasion.)** This information provides higher HQ with an idea of the evaders' initial actions or plans to assist in the preparation of a timely recovery effort.

    (a) ( ) Hiding intentions:

    (b) ( ) Evasion intentions: **(alone, buddy teams, split team, or whole team)**

    (c) ( ) Initial evasion point (IEP): **Explain actions at the point, and determine planned length of stay before moving.**

    (d) ( ) Travel intentions: **Address travel distances, duration of movements, time of day for movement, and so on.**

    (e) ( ) Handling of injured personnel: **Self-explanatory.**

    (5) ( ) Extended evasion goals:

    (a) ( ) Destination: **(border crossing, coastline, mountain range, recovery support area, and so on)**

(_____)
CLASSIFICATION

FM 3-05.231

(_____)
CLASSIFICATION
ANNEX W (EVASION AND RECOVERY) to OPERATIONS ORDER or PLAN (Number or Code Name), JSOA XXXXX, Co A, 2d Bn, 9th SFG(A)

    (b) ( ) Routes/movement corridors:

    (c) ( ) PLs: **Include the ETA and ETD for each PL. Ensure they are easily identifiable on the EPA Overlay.**

    (d) ( ) Evasion movement procedures:

    (e) ( ) Communications plan and radio procedures:

    (f) ( ) Signal plan:

    (g) ( ) HU sites and procedures:

    (h ( ) RV points and procedures:

    (i) ( ) Actions at the final evasion point (FEP):

    (j) ( ) RAS, load, far and near recognition signals:

    (k) ( ) Actions and procedures at potential contact points:

    (l) ( ) Border crossing procedures:

    (m) ( ) Coastline or marine operations procedures:

    (n) ( ) Recovery procedures:

    (o) ( ) Handling of injured personnel:

  (6) ( ) Evasion corridor boundaries: **Describe directions, distances, and linear features designated as boundaries. Use roads, rivers, and other features easily recognizable from the air, while on the ground, and on a map.** (Refer to Appendix 1, EPA Overlay).

  b. ( ) Other Missions (Subunit Tasks):

    (1) ( ) Specific instructions for subunit tasks:

    (2) ( ) Tactics: **IAW SOP. Refer to all tactical tasks specific to E&R and not addressed in OPORD or SFODA SOP.**

    (3) ( ) Actions for the care of sick and wounded: **Self-explanatory.**

    (4) ( ) Actions on civilian contact: **Describe what actions the detachment will take upon chance contact with civilians. Also address actions to be taken should contact with civilians be required; for example, medical assistance, acts of mercy, life or death decisions, and so on. Address breaking contact, detention, and so on.**

    (5) ( ) Actions at border crossings: **Cover actions to be taken, including the surveillance plan, mechanics of the crossing, and the selection of tentative crossing sites.**

    (6) ( ) Imminent capture procedures: **Detail procedures to be followed, including document control, equipment destruction, use of weapons, resistance posture, and so on.**

    (7) ( ) Resistance to interrogation and exploitation: **Address what actions to take while detained (for example, adhering to the Code of Conduct, sticking to a story, stalling, and so on).**

(_____)
CLASSIFICATION

(_____)
CLASSIFICATION
ANNEX W (EVASION AND RECOVERY) to OPERATIONS ORDER or PLAN (Number or Code name), JSOA XXXXX, Co A, 2d Bn, 9th SFG(A)

(8) ( ) Actions for escape: **Include subparagraphs for initial capture and movement to permanent facility and during confinement (organization, planning, communications, defeating control measures, and so on).**

c. ( ) Coordinating Instructions and Emergency Destruction Plan:

(1) ( ) State when EPA is effective: **Example: This EPA is effective upon receipt and approval by FOB or on order of FOB or senior coherent SFODA member.**

(2) ( ) Method used to synchronize the E&R time window: **DTG will be known as E day, H hour. E day will correspond to calendar day.**

(3) ( ) EPA review date (SFODA): **DTG.**

(4) ( ) EPA review date (with aircraft commander during aircrew mission brief): **DTG.**

4. ( ) SERVICE AND SUPPORT:

a. ( ) Equipment: **Cover all team equipment, equipment common to all, evasion aids, survival kit items, and special equipment to be used for E&R (for example, survival kit items, recovery equipment, beacons, and communications equipment). List the equipment here or in an additional appendix.**

b. ( ) Resupply:

(1) ( ) Aerial resupply: **(List separate paragraphs for on-call or automatic resupply, if applicable.) Cover locations, descriptions, time windows, and actions at resupply drop zone (RDZ) (if not addressed in paragraph 3). List contents here (or in Appendix 4, Logistics).**

(2) ( ) Cache: (Refer to Appendix 1, EPA Overlay, and Appendix 4, Logistics.) **Cover locations, descriptions, markings, time windows, and actions at cache site (if not addressed in paragraph 3). List cache contents here, or provide UNDER report as a tab to Appendix 4, Logistics.**

5. ( ) COMMAND AND SIGNAL:

a. ( ) Command: **List team in order of seniority. Include each team member's rank, name, SSN, DOB, duty position, call sign and code word, PLS number, blood chit number, and SERE training background (for example, Level B or C, SV-91/93, any advanced SERE training, and dates of training). List this information here or in subparagraphs.**

b. ( ) Signal:

(1) ( ) Communications:

(a) ( ) Code word from SFOB to SFODA to initiate evasion: **Verbal signal directing evasion.**

(_____)
CLASSIFICATION

FM 3-05.231

(_____)
CLASSIFICATION

ANNEX W (EVASION AND RECOVERY) to OPERATIONS ORDER or PLAN (Number or Code Name), JSOA XXXXX, Co A, 2d Bn, 9th SFG(A)

(b) ( ) Code word from SFODA to SFOB that evasion has been initiated: **Verbal signal confirming SFODA has initiated evasion.**

(c) ( ) Internal signals: **List all evasion signals to be used by the SFODA that are not covered by SOP or in the OPORD. Include team's evasion notification signals. These include verbal and visual signals for initiation, acknowledgement, and cancellation.**

(d) ( ) Radios, call signs, and frequencies: **Identify equipment, frequencies, and call signs dedicated specifically to the EPA. Be sure to specify number and type of survival radios and number of extra batteries. Use pertinent CSAR SPINS information here.**

(e) ( ) Beacon and SARSAT procedures: **Address beacon frequencies, frequency of transmission, and length of time beacon will be activated. Determine azimuth, elevation, and optimum time window for use of the SARSAT. Much of this will be dictated in the CSAR SPINS.**

(f) ( ) PLS codes: **Use last six digits of radio serial number(s) if not provided in paragraphs above. Include encryption code if used.**

(2) ( ) Other signals or special signals:

(a) ( ) GAS: **List day, night, primary, and alternate signals for aircraft extraction or aerial resupply.**

(b) ( ) RAS: **List day, night, primary, and alternate signals for aircraft extraction or aerial resupply.**

(c) ( ) Contact signals: **List day, night, alternate, and primary signals. May be included in CSAR SPINS.**

    1) ( ) Load signal:

    2) ( ) Far recognition signal:

    3) ( ) Near recognition signal:

(d) ( ) Straggler signals: **An inconspicuous but easily recognizable signal left at a site to indicate to stragglers that the site is currently occupied or has been previously occupied by other SFODA members. An additional signal to indicate the number of team members who have passed through the site should be selected. These two signals should be incorporated into the SFODA SOP.**

(e) ( ) LZ and DZ markings: **Include day and night and primary and alternate markings for each.**

    1) ERDZ: **Primary, alternate, day, and night markings.**

    2) Recovery HLZ or PZ: **Primary, alternate, day, and night markings.**

    3) Exfiltration PZ: **Primary, alternate, day, and night markings.**

(3) ( ) Authentication procedures: **Found in the ATO CSAR SPINS.**

(a) ( ) Number of the day:

(b) ( ) Letter of the day:

(c) ( ) Word of the day:

(d) ( ) Color of the day:

(_____)
CLASSIFICATION

FM 3-05.231

(_____)
CLASSIFICATION
ANNEX W (EVASION AND RECOVERY) to OPERATIONS ORDER or PLAN (Number or Code name), JSOA XXXXX, Co A, 2d Bn, 9th SFG(A)

    (e) ( ) Duress code word:

    (f) ( ) Signal code indicators:

    (g) ( ) SARNEG:

    (h) ( ) SARDOT:

  (4) ( ) Code words: **List any additional mission-specific or EPA-related code words here. Use five-letter words, all starting with the same letter, which represent locations, personnel, or actions.**

ACKNOWLEDGE:

                        CDR LAST NAME
                        RANK

OFFICIAL:

DATE PREPARED:
PREPARER'S LAST NAME AND RANK:
DUTY TITLE: **(for example, Operations Sergeant)**

APPENDIXES: 1. EPA Overlay

                2. Evasion RFAs for Hide Sites or Hole-Up Sites

                3. Contact Procedures and Linkup Plans

                4. Logistics **(resupply and cache bundle contents)**

                5. ISOPREP Cards

                6. EPA Cover Sheet

DISTRIBUTION:

(_____)
CLASSIFICATION

Appendix B

Forward Operational Base Evasion Plan of Action Guidance Format

The EPA of an SFODA is of prime importance to the overall successful accomplishment of its mission. Throughout the history of SF, plans for E&R have been a part of every mission. Many great examples of success and well-designed formats exist, but standardization has never occurred. EPA formats, procedures, and SOPs vary considerably from theater to theater, SF group to SF group, and SFODA to SFODA. The format on pages B-2 through B-8 provides the SFOB or FOB staff with a uniform standard for providing the commander's E&R guidance to the isolated SFODA. The format does not include examples of the FOB EPA appendixes.

INTENT

B-1. The intent of this appendix is not to instruct, but to provide the staff with a comprehensive format to assist the SFODA during mission planning. The overall objective is to institute a standard EPA guidance format that will streamline EPA planning, yet ensure continuity across the spectrum of SOF operations.

SF PR COORDINATOR

B-2. The SF PR coordinator or PR cell will take all available PR data and incorporate it into this five-paragraph OPORD format. This "incomplete" format will then be provided to the SFODA at the staff mission brief. The SFODA will take it back to the planning bay and incorporate its tactical mission plan and team-specific data. RFIs will be generated, as necessary. The SFODA will complete the EPA format and submit it at the briefback.

NOTE: Bold text in the following format example identifies those portions of the commander's EPA guidance that are unit- and/or mission-specific. These bold items also allow for the incorporation of the SFODA SOPs, where appropriate.

FM 3-05.231

(_____)
CLASSIFICATION

Annex W (Evasion Plan of Action OPLAN) to OPORD XXXX

References: **Map sheet (scale, major geographic area, sheet name and number, and series and edition), manuals, ARs, and other documents used to prepare the order or plan must be available to the EPA planner or supplied with the order. The order or plan of the next higher HQ is NOT a reference. A SAID is a reference. Show the references in block type.**

Time zone used throughout the order or plan: **Spelled out phonetically, in capital letters (for example, ZULU).**

Task organization: **Self-explanatory. Include the numbers for infiltration platform and personnel if known.**

1. ( ) SITUATION:

    a. ( ) Enemy Forces: See Intel Annex to OPORD **XXXX. Expand enemy situation throughout the E&R boundaries, to include population control measures; capabilities of police, constabulary, or militia; and demographics, to include attitudes of civilians (that is, anti-U.S., neutral, or pro-U.S.). If U.S. evading forces are seen by the local populace, what is their most probable COA?**

    b. ( ) Friendly Forces:

        (1) ( ) JSOTF PR Plan: **Example: JSOTF provides C3I on E day, H hour in support of SFODA XXX Evasion Plan of Action. OPLAN covers all areas within JSOA XXXXX to predesignated recovery points (PRPs), to facilitate a safe and timely exfiltration or emergency extraction of the detachment from the JSOA or specified boundary, allowing detachment and recovery force to return to FOB-92, or area controlled by friendly forces. Recovery forces can then rest and refit for follow-on employment.**

        (2) ( ) Friendly units to assist the SFODA during evasion:

            (a) ( ) **List all air assets providing infiltration and exfiltration platforms and resupply.**

            (b) ( ) **List theater asset responsible for providing on-call PR or CSAR support to FOB-92 during (DTG) (for example, JSOTF JSOAC).**

            (c) ( ) **List potential FFU elements to be encountered during passage of lines. May be listed as TBD.**

        (3) ( ) Fire and close air support available during evasion: **F-16, A-10, opportune, and on-call coordinated by JSOTF JSOAC. CH-47 or UH-60 and PR or CSAR element provide machine gun, light machine gun, and small-arms support. Reference Appendix 4 (Evasion RFAs for Hide Sites or Hole-Up Sites).**

    c. ( ) Attachments and Detachments: **As prescribed in OPORD XXXX and Bn Annex W (Evasion Plan of Action OPLAN) task organization.**

    d. ( ) Assumptions:

CLASSIFICATION

B-2

(_____)
CLASSIFICATION

Annex W (Evasion Plan of Action OPLAN) to OPORD XXXX

(1) ( ) **FOB-92** assumes the detachment has initiated its EPA if any of the following conditions exist:

Examples:

(a) ( ) **FOB-92 or the detachment transmits and receives acknowledgment of appropriate notification code word or phrase (see paragraph 5, COMMAND AND SIGNAL).**

(b) ( ) **FOB-92 has not received communications from the detachment within 24 hours after JSOAC confirms delivery of the detachment emergency resupply.**

(c) ( ) **The detachment misses primary, alternate, contingency, and emergency exfiltrations, and has not made subsequent communications with FOB-82.**

(2) ( ) If **FOB-92** has not received communications from the detachment within any given **24-hour period** (commencing on the DTG of infiltration), assume **FOB-92 has "triggered"** the MSR for delivery of the detachment emergency resupply or recovery. Triggering the MSR is the FOB response for a worst-case scenario, but does not signify activation of the detachment EPA. Reference Tab in Appendix 1 (Evasion Plan of Action Boundary Overlay) to Annex W (Evasion Plan of Action OPLAN) to OPORD XXXX.

(3) ( ) If **FOB-92** receives communications from the detachment **within 48 hours** of infiltration or last successful communication, assume **FOB-92** has **canceled** the MSR for delivery of emergency resupply or recovery bundle. Continue on with the assigned mission.

(4) ( ) If **FOB-92** has not received communications from the detachment within any given **72-hour** period, assume **FOB-92** has **actually launched** emergency resupply.

(5) ( ) **FOB-92** assumes the detachment will try to establish communications as soon as possible after recovery of the emergency resupply bundle.

2. MISSION:

a. ( ) **C3I:** Example: FOB-92 provides C3I on E day, H hour in support of SFODA XXX Evasion Plan of Action to facilitate a safe and timely exfiltration of the detachment to friendly lines or area controlled by friendly forces.

b. ( ) Commander's Intent: **Example: Enable the safe and timely exfiltration or emergency extraction of the detachment from the JSOA or specified area and return to FOB-92 or area controlled by friendly forces to rest and refit the detachment for follow-on employment.**

3. EXECUTION:

a. ( ) Concept of the Operation: See Appendix 1 (Evasion Plan of Action Boundary Overlay). Upon activation of EPA, **SFODA XXX** evades with attachments, if required, and conducts recovery or emergency resupply operations, using the evasion corridor that exits the **JSOA** to the **southwest to west**. This corridor leads the team generally to the **west** to conduct recovery or

(_____)
CLASSIFICATION

FM 3-05.231

(_____)
CLASSIFICATION
Annex W (Evasion Plan of Action OPLAN) to OPORD XXXX

linkup with an FFU, and eventual movement of the detachment back to **FOB-92**. See Appendix 1, Evasion Plan of Action Boundary Overlay.

    b. ( ) Subunit Missions:

        (1) ( ) OPCEN:

            (a) ( ) Establish liaison with (joint or combined) operations centers for air assets and FFUs, whose area of influence is within the E&R boundary. **(FFU may be a multinational force.)**

            (b) ( ) Coordinate (joint or combined) (aerial, maritime, and/or vehicular) recovery operations within the AO of the FOB.

            (c) ( ) Coordinate EPA boundaries/corridors supportable by (joint or combined) friendly aviation overflights. Ensure they support undetected overland movement out of the **JSOA**.

            (d) ( ) **FOB-92** coordinates contact point(s) and procedures for linkup with FFU.

            (e) ( ) Coordinate, prepare, and issue contact point(s) and procedures for linkup with FFU, as applicable, NLT 24 hours before the **EALT**.

        (2) ( ) Military intelligence detachment (MID): Obtain and issue intelligence updates within detachment EPA boundaries.

        (3) ( ) CA detachment: Obtain and issue CA updates for detachment EPA boundary.

        (4) ( ) SUPCEN:

            (a) ( ) Stage and coordinate delivery of emergency resupply bundles to the departure airfield, as required.

            (b) ( ) Emergency resupply materials must be placed into man-portable packages; for example, duffel bags or rucksacks (approximately 45 pounds each) before bundle rigging.

        (5) ( ) SIGCEN: Be prepared to load transponder frequency code(s) into survival radios (AN/PRC-112, CSEL, and so on) during the detailed communications briefing.

        (6) ( ) **SFODA XXX:**

            (a) ( ) Submit an initial draft EPA through the LNO to the OPCEN **within 48 hours** of the staff mission briefing. The FOB will staff the draft EPA, identify any issues or "show stoppers," and return it ASAP with recommendations.

            (b) ( ) Plan for PR or CSAR within the **JSOA** and/or within the designated EPA boundaries.

            (c) ( ) **Include the following enclosures with detachment EPA:**

                1) ( ) Appendix 1 (Evasion Overlay): Ensure all applicable information is **drawn directly onto the map sheet(s).** (See items (6)(d), (e), and (f) below).

                2) ( ) Appendix 2 (EPA RFAs for Hide Sites and Hole-Up Sites).

                3) ( ) Appendix 3 (Contact Procedures and Linkup Plans).

                4) ( ) Appendix 4 (Logistics).

                5) ( ) Appendix 5 (ISOPREP cards).

(_____)
CLASSIFICATION

FM 3-05.231

(_____)
CLASSIFICATION

Annex W (Evasion Plan of Action OPLAN) to OPORD XXXX

      6) ( ) Appendix 6 (EPA Cover Sheet).

     (d) ( ) Show proposed movement route(s), designated PLs, IEP, PRP, HU, and HS on Appendix 1 (Evasion Overlay).

     (e) ( ) Estimate distances detachment will **tactically** move during each **24**-hour period after activation of EPA.

     (f) ( ) Annotate **proposed 24-hour movement** PLs in Appendix 1 (Evasion Overlay) using PACE. PACE will cover from 24 to 72 hours after initiation of EPA and includes the following (if the terrain analysis allows a full PACE):

        1) ( ) **P**rimary exfiltration location.

        2) ( ) **A**lternate exfiltration location.

        3) ( ) **C**ontingency exfiltration location.

        4) ( ) **E**mergency resupply drop zone.

     (g) ( ) JSOTF JSOAC PR or CSAR element will contact the SFODA, with or without radios, per the detachment Annex W, Appendix 3 (Contact Procedures and Linkup Plans).

     (h) ( ) **FFU** designation: **If applicable,** the **FFU** will contact the SFODA, with or without radios, per the SFODA EPA, Appendix 3 (Contact Procedures and Linkup Plans).

  d. ( ) Coordinating Instructions:

    (1) ( ) Detachment EPA is effective upon receipt and approval by **FOB-92** for execution or on order by the FOB commander or detachment commander (senior ranking team member who is coherent and uninjured).

    (2) ( ) The DTG the detachment EPA is activated will be known as **E day, H hour**.

    (3) ( ) The emergency resupply MSR trigger begins on the **24th hour after the last missed communication or infiltration TOT if no communication has been received by FOB-92**.

    (4) ( ) If the aircraft goes down in enemy territory during either infiltration or exfiltration, remain with the surviving aircrew members, and determine the best COA (to include the aircrew's existing ERPA). If radio communications are not possible with the FOB, the senior ranking coherent SFODA member has the authority to make the decision according to the best COA and METT-TC.

4. ( ) SERVICE SUPPORT.

  a. ( ) Submit a request for emergency resupply items to the SUPCEN. Include a copy in Appendix 4 (Logistics).

  b. ( ) Use the following destruction measures for equipment and/or sensitive components onboard aircraft:

    (1) ( ) Tactical destruction measures:

      (a) ( ) PRI: Thermite grenades.

      (b) ( ) ALT: Dismantle and bury sensitive items on aircraft.

    (2) ( ) Overt destruction measures:

(_____)
CLASSIFICATION

FM 3-05.231

(_____)
CLASSIFICATION
Annex W (Evasion Plan of Action OPLAN) to OPORD XXXX

    (a) ( ) PRI: Explosives. Use nonelectric, time-delay firing system.

    (b) ( ) ALT: Small arms.

5. ( ) COMMAND AND SIGNAL.

  a. ( ) Command:

    (1) ( ) Detachment chain of command differs only in that the **senior ranking coherent SFODA member** will be in **command**.

    (2) ( ) **Aircrew** falls beneath the detachment chain of command during evasion. Military courtesy will be maintained between junior and senior pay grades.

    (3) ( ) After linkup with FFU (or friendly asset), detachment falls beneath the existing chain of command of the FFU until returned to the control of the FOB.

  b. ( ) Signal:

    (1) ( ) Current SOI is in effect.

    (2) ( ) Team code name: **XXXXX**.

    (3) ( ) Primary and alternate means by which the FOB and/or ODA will initiate E&R:

      (a) ( ) **E&R Alert**: **NOTE** is the primary and alternate means of alerting the **FOB** or **SFODA** that the EPA may **soon** be triggered.

        1) ( ) The primary means of alert will automatically occur after any of the events take place in the assumptions paragraph. (See paragraph 1.d., Assumptions.)

        2) ( ) The alternate means of alert is the transmission of the encrypted phrase "**EVASION ALERT**" over the primary or alternate means of communication.

      (b) ( ) EPA Notification:

        1) ( ) PRI (encrypted): **ACTIVATE EVASION PLAN (Effective DTG).** This phrase triggers activation of detachment EPA.

        2) ( ) ALT (plaintext code word or phrase): **XXXXX** used in any plaintext combination. **NOTE:** The alternate code word or phrase will be used only if encryption is not possible (cryptography is lost or compromised).

      (c) ( ) Acknowledgment: Upon receipt of EPA notification, the receiving station must acknowledge as soon as possible.

        1) ( ) PRI (encrypted): "**ROGER, ACTIVATE EPA (EFFECTIVE DTG), OVER.**"

        2) ( ) ALT (plaintext code word or phrase): "**ROGER, COPY "XXXXX," OVER**," used in any plaintext transmission. **NOTE:** The alternate code word or phrase will be used only if encryption is not possible (cryptography is lost or compromised).

      (d) ( ) Duress code word or phrase (**MESSAGE TRAFFIC**): The **absence** of the code word or phrase "**XXXXX**" in message traffic notifies **FOB-92** that the detachment **is or was operating under duress** at the time of transmission. **FOB-92** will then initiate the EPA for the SFODA.

CLASSIFICATION

B-6

_____ FM 3-05.231

(_____)
CLASSIFICATION

Annex W (Evasion Plan of Action OPLAN) to OPORD XXXX

(e) ( ) Duress code word or phrase (**VOICE TRAFFIC**): The **presence** of the code word or phrase "**XXXXX**" in a voice message notifies **FOB-92** that the detachment **is or was operating under duress** at the time of transmission. **FOB-92** will then initiate the EPA for SFODA.

c. ( ) Communication procedures:

(1) ( ) Bring the detachment survival radios (AN/PRC-112, CSEL, and so on) to the detailed communications briefing. Radios will receive from the SIGCEN a transponder code that is unique to the detachment. This allows the aircrew and PR or CSAR element to authenticate and positively identify the detachment on the ground. Include the transponder code (**full code to include zeros**) in the detachment EPA in Annex W, Paragraph 5 (COMMAND AND SIGNAL); in Appendix 3 (Contact Procedures and Linkup Plans); and in Appendix 6 (EPA Cover Sheet).

(2) ( ) Timeline for making contact with friendly aircraft overflights: To prolong battery life, **simply turn on the survival radio for a maximum of 10 minutes every third hour. The cycle begins at midnight using JSOA local time (for example, 2400, 0300, 0600, 0900, 1200, 1500, 1800, and 2100). PR or CSAR assets will attempt recovery during one of these windows.**

(3) ( ) The primary ground-to-air emergency communication system is the **AN/PRC-112 beacon. Transmit on beacon setting for a maximum of 10 minutes followed by monitoring the AN/PRC-112 for 10 minutes. Continue as described in item (2) above, or after positive visual identification of friendly aircraft.** This includes the MH-53, MH-47E, and MH-60. **Follow all instructions from the pilot.**

(4) ( ) Friendly aircraft can interrogate the AN/PRC-112 in the passive mode (AN/PRC-112 turned on), authenticate the transponder code, and locate the position of the evader's radio. **NOTE:** For training, do not use the beacon transmitter unless a real-world emergency exists.

(5) ( ) Develop an alternate air-to-ground communications plan according to the communications capabilities of the mission (FM, UHF, or VHF). Prepare this plan with the assistance of the SIGCEN during the detailed communications briefing. Include emergency ground-to-air frequencies and the proposed team call sign in the detachment EPA. **NOTE:** Keep the call sign for the team simple.

(6) ( ) Develop a day-night HLZ visual marking authentication system (for example, panels, mirror, strobe, chemical light, or pen-gun flares). **NOTE:** Keep the system simple.

(7) ( ) ATO SPINS for day/month/year: **DAY 01 MONTH APR YEAR 00**

  (a) ( ) LETTER: **U.**

  (b) ( ) WORD: **SHOVEL.**

  (c) ( ) COLOR: **BLACK.**

  (d) ( ) NUMBER: **27.**

  (e) ( ) DURESS WORD: **COORS.**

(8) ( ) Other signaling criteria:

(_____)
CLASSIFICATION

FM 3-05.231

(_____)
CLASSIFICATION
Annex W (Evasion Plan of Action OPLAN) to OPORD XXXX

    (a) ( ) MIRROR: any auto (for example, CORVETTE).

    (b) ( ) STROBE: any beer (for example, BUDWEISER).

    (c) ( ) SMOKE: any cigarette (for example, MARLBORO).

    (d) ( ) Expedient single transposition cipher (to transmit grid coordinates, and so on):

```
        O R A N G E P L U M
        0 1 2 3 4 5 6 7 8 9
```

    (e) ( ) RAS:

        1) ( ) READY FOR PICKUP: Ü

        2) ( ) STILL MOVING: U

    (9) ( ) All radio frequencies and code(s) are: **REQUIRED** in the EPA (**full code, to include zeros**) in Paragraph 5, COMMAND AND SIGNAL; in Appendix 3 (Contact Procedures and Linkup Plans); and in Appendix 6 (EPA Cover Sheet).

NOTE: Radio frequencies are received during battalion communications brief.

    (10) ( ) GAS: **Day-night signals that can be used anytime, in the absence of communications, to signal aircraft identified as friendly. Display at 3-hour intervals (2400, 0300, 0600, 0900, 1200, 1500, 1800, and 2100 local time) for a maximum of 10 minutes after the specified hour.**

Acknowledge

                                IMA A. HERO
                                LTC
                                Commanding

OFFICIAL:
SMITH
CURRENT OPERATIONS OIC

Distribution:
FOB-92 OPCEN
Team **SFODA XXX**

Appendixes:

1. Evasion Plan of Action Boundary Overlay

2. Weather Data

3. List of Operational Evasion Charts Published by DMA

4. Evasion RFAs for Hide Sites or Hole-Up Sites (**NOTE:** Completed by SFODA.)

5. Contact Procedures and Linkup Plans (**NOTE:** Completed by SFODA.)

6. Evasion Plan of Action Cover Sheet (**NOTE:** Completed by SFODA.)

(_____)
CLASSIFICATION

Appendix C

Evasion Plan of Action Planning Considerations

During times of conflict, SF performs missions at the strategic and operational level to influence deep, close, or rear operations. SF is optimally designed to conduct and support theater deep operations beyond the forward limits of conventional land forces. Such operations may extend deep into the homeland of a hostile power or into the territory of hostile states in pursuit of theater strategic military objectives. Often clandestine or covert in nature, these operations are conducted across the full range of military operations, frequently independent from friendly support, and characterized by high political and physical risk. SF personnel are subject to isolation in hostile territory and should be prepared for the possibility of finding themselves in an evasion situation. Successful evasion is dependent on effective prior planning. The development of an EPA should be a part of every SO mission planning effort.

The EPA is a COA developed before executing a combat mission. The EPA is intended to improve a potential evader's chances of successful evasion and recovery by providing recovery forces with an additional source of information that can increase the predictability of the evader's actions and movement (JP 3-50.3). The EPA is one of the critical documents for successful recovery planning. The EPA is a vehicle by which potential evaders, before isolation in hostile territory, can plan their potential evasion in detail and relay those intentions to a recovery force. Potential evaders, with the aid of intelligence personnel and PR planners, should complete the EPA based on thorough knowledge of the environment where isolation may occur. Evaders may gain such knowledge by studying the combat environment and the hostile territory during the predeployment phase of a special operation. When possible, the EPA should be coordinated with the recovery elements.

EVASION PLANNING RESOURCES

C-1. The SFODA has multiple evasion planning resources available in most circumstances. The sources in the following paragraphs are produced specifically for PR and evasion planning. Additional awareness and insight can be gained from other sources not specifically produced for PR, such as noncombatant evacuation plans, U.S. Marine Corps Country Handbooks, or operational support studies.

PR APPENDIX TO THE THEATER OPLAN

C-2. The OPLAN PR appendix outlines the theater's basic concept of operations for PR. This appendix also assigns tasks for the recovery of isolated personnel from enemy controlled or hostile territory.

CONOPS OR SEARCH AND RESCUE STANDING OPERATING PROCEDURES

C-3. The CSAR CONOPS or search and rescue standing operating procedure (SARSOP) describes the overall theater concept for integration and coordination of available capabilities to accomplish the five phases of PR (report, locate, authenticate, support, and recover). The PR CONOPS addresses situations ranging from rear area SAR conducted by the HN to a combat recovery mission involving dedicated recovery assets and NAR.

CSAR SPINS

C-4. CSAR SPINS are part of the ATO and/or ITO and provide theater guidance for PR operations. CSAR SPINS are specific to the theater of operations, and their content will vary from one AO to the next. They are normally updated on a weekly and quarterly basis or as needed if compromised. They will often address theater-wide authentication codes and procedures, an isolated person's legal status, communications schedule and procedures, extended evasion procedures, and no radio (NORDO) signal procedures. When available, CSAR SPINS should be incorporated into the SFODA EPA to the fullest extent possible to maximize their chance of successful recovery.

DAILY CODES

C-5. Daily codes include the number, word, and letter of the day. Each of these or a combination of these can be used as a means of authentication. These daily codes are available to everyone and articulated in the ATO and/or ITO. These codes, therefore, provide on-scene aircraft the means to rapidly authenticate isolated personnel. SFODAs must indicate in their EPA the daily codes in effect during the day of infiltration. All codes indicated on their EPA will remain in effect for the duration of their mission.

DURESS WORD

C-6. The isolated person indicates to recovery forces his transmission is being forced. The isolated person indicates this by using the duress word provided in the CSAR SPINS.

COMMUNICATIONS PROCEDURES

C-7. The CSAR SPINS normally provide a communications schedule and assign specific frequencies to isolated personnel and recovery forces. Specific procedures for the use and programming of survival radios, including the personal locator system (PLS), and radio identification codes are routinely provided.

RAS

C-8. A RAS is a precoordinated signal from an evader that indicates his presence in an area to a receiving or observing source. A RAS indicates, "I am here, start the recovery planning." The JSRC may select a theater-wide RAS and include it in the CSAR SPINS. The RAS may be a GAS, transmitted via radio, or another nontechnical communications signal.

NORDO

C-9. NORDO procedures outline the use of specified GASs when technical communications are not available or are not tactically feasible. A GAS could include the use of infrared lights or symbols displayed on the ground. This description should include any intended offset, specific orientation of the signal, or any intentions evaders and recovery forces wish to indicate with the signal.

SARDOT

C-10. A SARDOT is a geographical location known only to friendly forces, which allows an isolated individual to pass his location over an unsecured radio net. The CSAR SPINS or CONOPS will normally direct coordinate format, map datum, and GPS programming procedures, including magnetic bearing and nautical miles. A SARDOT is usually given an easily remembered code word and is used by transmitting the range and bearing to the code word; for example, **"Viper 25 is 325 for 20 to Daffy."**

SARNEG

C-11. The SARNEG is a 10-letter code word with no repeated letters that corresponds to the numbers 0 through 9. It allows an isolated individual to pass his encrypted location over an unsecured radio net. Normal procedures are to pass latitude then longitude without regard for the directional indicators. An example of a SARNEG code is shown below.

SARNEG	C	O	M	B	A	T	H	E	L	P
TO ENCODE	0	1	2	3	4	5	6	7	8	9

26 30N 011 25 E is passed in SARNEG as follows: **MHBCCOOMT**

INTELLIGENCE

C-12. Before formulating the EPA, SFODA members should coordinate with unit intelligence personnel. The minimum essential elements of information (EEI) include the following:

- Air, ground, electronic, and naval order of battle, to include friendly and enemy troop dispositions, capabilities, and estimated response times.
- Climatic conditions and extremes.
- Terrain, flora, fauna, cover, and concealment.
- Indigenous groups' identifiable dress, habits, or customs that will aid the isolated personnel in identifying friendly and hostile elements.

- Population control measures limiting or restricting movement, such as checkpoints and roadblocks.
- Predetermined recovery areas.
- Identifying characteristics of international boundaries.
- ROE.
- Legal status.
- Neutral countries.
- Other EEI, as required.

JPRSP

C-13. The JPRSP is a country-wide, module-based, scalable digital product designed to meet geographic combatant commander PR requirements by placing collaborative-produced intelligence in a single location for use by PR planners, intelligence personnel, operators, aircrew members, rescue forces, and others needing the information. The JPRSP is the primary production tool and replacement product for all earlier PR intelligence products that no longer fully support operational requirements. The production of the JPRSP is prioritized according to the requirements and intelligence production capability of the geographic combatant commander. Previous E&R studies and products should be used until replaced by a JPRSP. The JPRSP will include information about threats to the evader and recovery forces; SERE procedures for survival, evasion, captivity, resistance, and escape; known detention facilities; geographical and terrain studies; survival skills; and other information found in earlier PR intelligence documents. PR planners should use the JPRSP in preparing the EPA and recovery plans.

PREDETERMINED RECOVERY AREAS

C-14. Nationally produced SAFE and DAR products are examples of predetermined recovery areas. SFODAs should nominate areas according to mission needs when nationally covered areas are unavailable or infeasible. Theater planners, units, and individual SFODAs may designate areas according to their mission requirements. They should coordinate with recovery forces and intelligence functions to ensure designated areas are supportable.

SAFE

C-15. A SAFE is a designated area in hostile territory that offers evaders or escapees a reasonable chance of avoiding capture and surviving until they can be evacuated. Selection criteria for a SAFE can be found in JP 3-50.3, Appendix G, which is published separately from the manual.

SAID

C-16. A SAID contains an in-depth, all-source evasion study designed to assist the recovery of military personnel from a selected area for evasion under hostile conditions. More detailed information concerning a SAID can be found in JP 3-50.3, Appendix G. SAIDs are generally developed via technical means, and human intelligence (HUMINT) resources do not validate information contained therein.

E&R AREA STUDIES

C-17. E&R areas may be selected in any geographic region according to operational or contingency planning requirements. Although similar to SAFE areas in most respects, E&R areas differ in that not all conventional selection criteria for SAFE areas can be met because of current political, military, or environmental factors prevailing in the country.

SERE GUIDES AND BULLETINS

C-18. These publications are essential reference documents for potential evaders. They contain the basic information to help an individual survive, successfully evade and, if captured, resist enemy exploitation. SERE guides and bulletins cover an entire country or region of the world and provide information on—

- Topography and hydrography.
- Food and water sources.
- Safe and dangerous plants and animals.
- Customs and cultures.
- Recognition of hostile forces.
- Resistance techniques in captivity.
- Other types of information.

In short, they include the type of general information that serves as the foundation upon which the more specific information found in EVCs, SAIDs, and current intelligence briefings can be used to build a sound evasion plan.

DD FORM 1833

C-19. As part of the mission planning process, SFODA members and appropriate aircrew members should review their DD Form 1833 (ISOPREP). When filled in, the ISOPREP is classified Confidential. It enables a recovery force to authenticate isolated personnel and ensure a successful recovery. The unit and the individual are responsible for ensuring that the ISOPREP card is properly completed, kept on file, and ready to be transmitted to the JSRC IAW established theater standards. Failure to accomplish this requirement complicates recovery planning, puts the recovery force at risk, and jeopardizes the success of the recovery mission. Joint doctrine outlines the minimum requirements for a completed ISOPREP. Theater-level PR planners routinely provide additional guidance on ISOPREP data, such as inclusion of boot and uniform sizes. Explicit instructions for completing the ISOPREP card are in the theater PR CONOPS, SARSOP, regulations, and CSAR SPINS.

SF PR COORDINATOR

C-20. The SFOB, FOB, or AOB PR coordinator provides information on theater PR plans and procedures. The PR coordinator is responsible for assisting isolated SFODAs in the development of their EPAs and reviewing **all** EPAs and ISOPREPs for completeness.

JSOTF AIR PLANNERS AND AIRCRAFT COMMANDER

C-21. The SFODA should coordinate with the air planners any actions affecting the EPA. They should consider actions to take if isolated with the air component during each planned leg of the mission.

CACHES

C-22. Caches may be pre-positioned in enemy-controlled territory or in regions subject to being overrun by enemy forces. Their use should be considered in environments where extended evasion is projected. Evaders can use caches as sources of supplies (food, water, clothing, and so on), communications equipment, weapons, and other evasion aids. In denied areas, unconventional assets may emplace caches before or after the outbreak of hostilities. Agencies and organizations that direct the establishment of caches that could potentially support evaders need to keep the JSRC advised of the status and locations of those caches to provide optimal support.

EVASION AIDS

C-23. Blood chits, pointee-talkees, EVCs, and other equipment should be obtained well in advance of operations. The Service or theater PR OPR is responsible for blood chit distribution before deployment, and the JSRC is the normal point of contact for the theater to obtain blood chits once deployed. EVCs are standard map products produced by NIMA and may be procured through the S-2. Additional evasion aids are developed via Service and theater requirements and procedures.

EN ROUTE EVASION PLAN

C-24. The en route evasion plan is devised and coordinated jointly by the SFODA and the infiltration aircrew with the assistance of JSOTF air planners and the SFOB or FOB OPCEN, and with the insertion plan for the mission. The SFODA should be provided with the base plan during the staff mission brief and complete the plan during isolation or mission planning. The isolated SFODA and the supporting infiltration aircrew must complete the en route evasion plan. This plan articulates the intentions of the SFODA and the infiltration aircrew should they have to initiate their EPA during the infiltration or exfiltration phase of the mission.

C-25. The en route evasion plan covers the entire route from the departure airfield, through friendly areas, to the initial penetration of enemy controlled territory and, ultimately, to the objective area. This plan may be divided into phases (for example, over water leg, overland leg in neutral or friendly territory, after penetration of hostile airspace or sea space, before and after the point of no return, and so on) with separate actions for each phase.

C-26. The en route evasion plan may be the same as the SFODA EPA if the SFODA evasion corridor includes the insertion route. For example, if the SFODA is downed after the MDL and rescue is not immediately available, the

SFODA will evade toward their designated evasion corridor and follow its EPA.

C-27. During situations in which recovery is not immediately available, the initial focus should be on actions for emergency bailout, downed aircraft procedures, movement away from the crash site or landing area, treatment of wounded, removal and destruction of sensitive equipment, and establishment of initial communications by using survival or standard SFODA radios. Once in a hide site or hole-up site, the SFODA may remain stationary to avoid detection or continue moving toward an ultimate evasion destination.

C-28. During helicopter insertions and extractions, the aviation package (if multiple aircraft) may be able to conduct emergency extraction of personnel, if necessary. Hot LZ or pick-up zone (PZ) procedures should be coordinated between the SFODA and the insertion aircrew. The en route plan is divided into two phases:

- *Phase I.* The first phase is from friendly areas to the MDL. In the first phase, the SFODA is not close enough to its objective to execute the mission. As a matter of policy, the senior-ranking SFODA survivor is in command. The SFODA should focus on maintaining the security of the mission, caring for wounded, surviving, self-initiating rescue (establishing communications), and evading.

- *Phase II.* The second phase is that portion beyond the MDL near enough to the objective for the SFODA to possibly continue its mission according to the original abort criteria. In some situations, the SFODA may execute its EPA once beyond the MDL, instead of the en route evasion plan. After long-range communications have been established, the senior-surviving SFODA member may decide that the SFODA can still move to the objective area in time to accomplish the mission. This decision will be made considering the level of compromise, the minimum force requirements, and abort criteria. The insertion crew may accompany the SFODA on its mission or evade to an area where rescue can occur.

C-29. The en route evasion plan is closely coordinated and integrated into the insertion plan. The en route evasion plan applies to the SFODA and all aircrew. The SFODA and aircrew must agree on basic evasion tactics, such as actions for movement and actions at recovery points. Normally, the aircrew will move in the center of the element, normal SFODA tactics and SOPs will be used, and the senior-surviving SFODA member will assume tactical control (TACON) of all personnel.

C-30. Normally, the SFODA and infiltration aircrew prepare the en route evasion plan by using EVCs and 1:250,000-scale maps and overlays to show the entire insertion route. The SFODA may also need to study 1:50,000-scale maps and imagery products to memorize key points, such as air routes, air checkpoints, initial penetration point, MDL, predetermined recovery points, and so on. The SFODA member preparing the en route evasion plan and the infiltration aircrew should coordinate the final insertion evasion plan during the air mission brief, since this is the first time the SFODA will see the flight route. They must agree on actions to be taken for downed aircraft, emergency bailout, abort criteria, alternate insertion plans, contingencies for insertion, and joint evasion. The CSAR SARDOT and CSAR SPINS information will be drawn from the ATO. The SFODA member conducting evasion planning should complete an en route evasion plan checklist. This checklist will be submitted to the OPCEN at the completion of the briefback. The infiltration aircrew should be provided a copy of this plan after the air mission brief.

EVASION CORRIDOR

C-31. The EPA applies to the SFODA for the duration of its assigned mission. The SFODA will normally be assigned a corridor from the JSOTF or SFOB OPCEN and provided with the SFOB or FOB evasion planning guidance during the staff mission brief. The SFODA completes the plan during isolation or mission planning. The corridor should be designed around suggested or predetermined recovery areas. The corridor should be planned along easily recognizable terrain features to provide for easier navigation, since potential evaders may have limited navigational aids (for example, small-scale maps). The corridor should not be so small that it limits maneuver room (for example, not less than a few kilometers wide), nor should it be so large that it makes searching difficult. The corridor may narrow when approaching the FEP, making the SFODA approach along a designated route, or from a specific direction (that is, a linkup), which may be a coordination measure designed to prevent fratricide. The corridor covers the general route the SFODA will take through an area where the SFODA may be rescued, may return to friendly territory via a passage of friendly lines, or make a border crossing. The corridor also acts as a control measure that will aid the OPCEN in tracking the movements of the SFODA during evasion and makes searches by recovery forces easier.

C-32. The evasion corridor extends from the objective area to the FEP. The FEP may be a location that will remain under friendly control for the duration of the mission or a designated recovery area within enemy territory where an evader can safely await recovery or linkup. The FEP is the ultimate evasion destination of the SFODA should they have to evade unassisted. The corridor may extend back toward a forward edge of the battle area (FEBA) (if one exists). However, this type of corridor should normally be avoided because of the heavier concentrations of troops, battlefield obstacles such as minefields, and the danger of fratricide. The corridor will normally extend to an FEP within a designated SAFE or to an area where the SFODA can survive, avoid detection, and be recovered.

C-33. Within the corridor, the SFODA will select a PRP location. It may be selected from the existing helicopter landing zones (HLZs) identified within a SAID for the AO. It is preferred that there be one PRP for each day of evasion anticipated. There may be no preplanned SAFEs (selected area for evasion) along the evasion route of the SFODA. In this case, the SFODA will select and designate additional PRPs for each day of anticipated evasion. The distance between PRPs will depend on hours of darkness available for movement and how far the SFODA can move within that time. The goal is for the OPCEN to track the movement of the SFODA during evasion to assist recovery forces in locating the SFODA for recovery.

C-34. Recovery planners use EVCs, maps of all scales, and overlays to plan the corridor. The SFODA will receive its maps and overlays with its FOB EPA guidance. If available, recovery planners should use current imagery and SAIDs to plan the corridor.

C-35. The SFODA should memorize the entire EPA by using a detailed map study, noting as much detail as possible given the time available. The SFODA will study 1:50,000-scale maps and imagery to memorize key points for evasion (for example, PRPs and SAFE locations, final evasion points, linkup sites, and so on). All SFODA members should be able to briefback the entire EPA. All SFODA members should be able to identify all EPA locations on an EVC or unmarked map.

OTHER CONSIDERATIONS

C-36. Successful recovery is dependent on effective prior planning. The responsibility ultimately rests with the SFODA. Each SFODA member must be familiar with all information contained in his EPA. The EPA, along with the ISOPREP, contains critical information to facilitate recovery. The EPA enables the recovery force to conduct a successful tactical recovery. The SFODA should execute its EPA as planned to ensure successful recovery. EPA and ISOPREP information must be readily available to PR planners and the recovery force. The EPA and ISOPREP should be maintained in the OPCEN to facilitate the passing of critical information to the appropriate PR command and control node.

NORDO

C-37. Although not directly part of the EPA, not making communications after insertion or loss of communications for a given time during a mission may be reason for the SFODA to begin evasion. The OPCEN and the SFODA must have a clear understanding of how not having or losing a communications plan ties in with the EPA; therefore, they should address the exact criteria for initiation of the EPA in the absence or loss of communications. Normally, an SFODA will be extracted by using recovery assets instead of being resupplied during a NORDO situation. Since the situation of the SFODA is unknown, the OPCEN will most likely assume the SFODA is compromised according to the multiple redundancies of modern SF communications. Resupply aircraft may increase the level of compromise or be unnecessarily placed at risk.

EMERGENCY DESTRUCTION OF EQUIPMENT

C-38. Each SFODA will develop a plan for the destruction of its communications equipment and classified materials in case capture is imminent or equipment has to be discarded. The SFODA will destroy survival radios and equipment and navigational aids only as a last resort.

PLAN FOR SURVIVAL

C-39. Each mission packet of the SFODA should contain the SERE contingency guide and the isolated personnel guidance (commonly referred to as the IPG) for the AOR. This information, combined with the weather forecast and situation briefings, will be used to formulate a plan for survival. Knowing how to survive will be extremely important during prolonged evasion or captivity.

Appendix D

Unconventional Assisted Recovery Coordination Center Operations

This appendix discusses the role of the UARCC in PR. It also discusses the manning and operations of the UARCC.

UARCC MISSION

D-1. The mission of the UARCC is to integrate and coordinate all theater NAR capabilities in support of the JFC's PR operations. The UARCC is a compartmented SOF facility staffed on a continuous basis by supervisory personnel and tactical planners to coordinate, synchronize, and deconflict NAR operations within the JFC's operational area. The UARCC interfaces and coordinates with the JOC, the JSRC, and when directed, with each component RCC. Once established, the UARCC conducts the following critical tasks:

- *Advises*. The UARCC advises the commander on the development and employment of NAR capabilities in support of the theater PR plan.
- *Communicates*. The UARCC provides the connectivity between the NAR forces and the theater PR architecture to provide time-critical information and acts as the conduit through which launch-and-execute authority is passed.
- *Coordinates*. The UARCC coordinates all representative NAR capabilities. It coordinates support for the comprehensive NAR ground tactical plan. When directed by the JSRC, it coordinates with the RCCs of the theater components for conventional recovery support.
- *Integrates*. The UARCC integrates NAR capability into the theater PR plan.
- *Deconflicts*. The UARCC deconflicts NAR operations internally with all NAR forces supporting a single recovery operation. It deconflicts NAR operations externally with other theater operations. The UARCC conducts timely exchange of operational and support information to aid mission execution, avoid disruption of ongoing operations, and avoid fratricide.
- *Synchronizes*. The UARCC synchronizes the ground tactical plans between NAR tactical elements. Synchronization also occurs between NAR tactical elements and conventional recovery support, conventional military operations, SO, and interagency activities.

UARCC COMMAND RELATIONSHIPS

D-2. The UARCC is the singular NAR facility in the theater and plays an integral role in the comprehensive theater Personnel Recovery Management

System. The UARCC is responsible for the coordination and management of theater NAR forces (Figure D-1). The following paragraphs discuss the command relationships of the JFC and JFSOCC and the launch and execute authority.

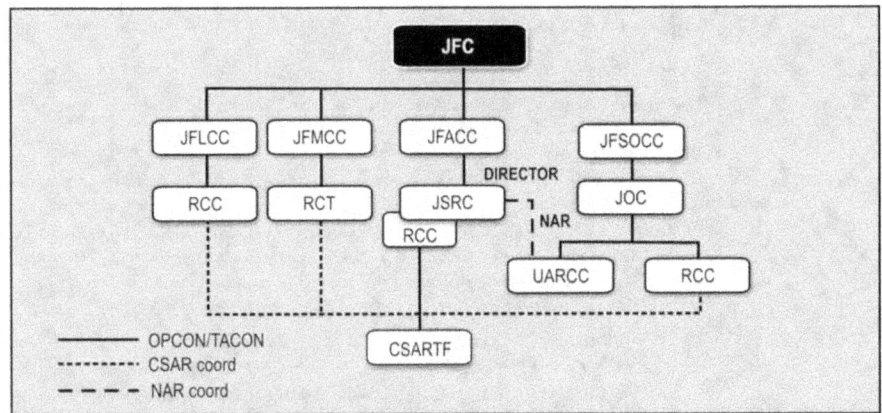

Figure D-1. UARCC In-Theater PR Architecture

JFC

D-3. The JFC is responsible for PR support of assigned forces. He is responsible for developing an integrated PR program incorporating conventional and nonconventional PR capabilities. JFCs normally exercise command authority for PR through the JFAAC via the JSRC. If the JFC chooses to designate a JSRC director as a member of his special staff, the JFC could centrally manage theater PR. The JSRC is charged to coordinate PR for the theater. The JSRC has the authority to coordinate support to the UARCC for NAR operations.

JFSOCC

D-4. The JFC designates the JFSOCC as the OPR for the planning, coordinating, and executing of all NAR activities in support of the theater PR plan. The JFSOCC retains command authority of all SOF UAR forces in-theater. The JFSOCC exercises C2 through the J-3, who designates a UARCC director and, on order, establishes the UARCC. The UARCC coordinates NAR activities but does not exercise command authority. The JSOTF and/or FOB retains C2 of UAR forces (SFODAs). OGAs supporting NAR normally retain C2 of their respective forces.

LAUNCH-AND-EXECUTE AUTHORITY

D-5. Command relationships for launch-and-execute authority must be clearly articulated before the commencement of NAR operations. The JFC may retain execute authority or delegate that authority to the JFACC or JFSOCC.

UARCC STAFF ROLES AND RESPONSIBILITIES

D-6. The branch of the SOC J-3 normally responsible for compartmented projects forms the nucleus of the UARCC. The UARCC should be staffed and prepared for continuous operations. The UARCC is located in a compartmented facility.

D-7. Once established, the UARCC is required to continuously operate until directed to stand down. PR responsibilities of the UARCC include, but are not limited to, the following:

- Maintaining direct and continuous liaison with tactical planners of designated NAR assets.
- Conducting mission planning and COA development and analysis. Making recommendations for NAR operations to the JFSOCC or his designated representative.
- Developing IPB.
- Establishing procedures to coordinate support to recovery forces and monitoring mission progress and status of recovery assets.
- Coordinating, integrating, and deconflicting interservice and interagency NAR joint TTP within the theater to contact, authenticate, support, move, and exfiltrate isolated personnel.
- Maintaining records containing all available data on personnel recovered via NAR.
- Forwarding all information acquired by NAR forces concerning status of missing, evading, or captured personnel to the JSRC.
- Coordinating with the JSRC to support the debriefing of personnel recovered via NAR by JPRA J-32.

UARCC MANNING

D-8. The UARCC staff is typically a director, one shift supervisor per shift, controllers (operations officers and NCOs), intelligence analysts, communications officers and NCOs, and representatives of tactical organizations possessing NAR capabilities (Figure D-2, page D-4). On order, SFOB or FOB commanders provide a tactical planning element, generally an operational control element (minus) (OCE[-]), to the UARCC to plan and coordinate ground tactical plans in which SFODAs are participants. These tactical planners must be fully aware of detailed capabilities of the ground tactical elements they represent. They must also possess the appropriate security clearances allowing them direct interface with all other NAR force representatives. These tactical planners maintain communications connectivity with their C2 nodes and maneuver forces. The tactical planners represent maneuver forces in comprehensive NAR planning, briefings, and mission execution. They monitor the status of deployed UARTs and UARMs. The tactical planners maintain visibility of their respective maneuver elements and are prepared to brief, as appropriate.

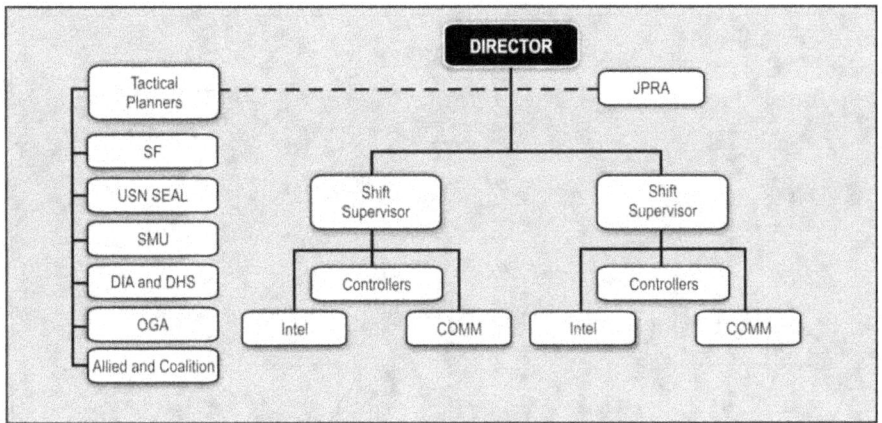

Figure D-2. Typical UARCC Organization

UARCC OPERATIONS

D-9. Mission planning usually begins when the UARCC is notified that an individual is missing or isolated. Notification is generally via SARIR, SARSIT, or an observation report of RASs.

PARALLEL PLANNING

D-10. Once notified of a PR incident, the UARCC immediately begins parallel planning, as do all other RCCs. The UARCC conducts a capability and feasibility assessment. Upon completion of the assessment process, the UARCC informs the JSRC if NAR capabilities exist. The UARCC continues planning until such time the JSRC informs the UARCC its assets are not required or a CONOPS is developed. The UAR planner uses a PR planning matrix (Figure D-3, page D-5) to aid him in determining UAR capabilities, limitations, and requirements.

COA DEVELOPMENT

D-11. IAW the commander's planning guidance, the tactical planners conduct their capability and feasibility assessments. The tactical planners consider the following:

- Time and access.
- Risk versus gain.
- Capacity and movement.
- Capability and limitations.
- Threat and current intelligence.
- Evader EPA.
- Applicable ATO or SPINS information specific to evader incident; for example, word, letter, and number of the day.

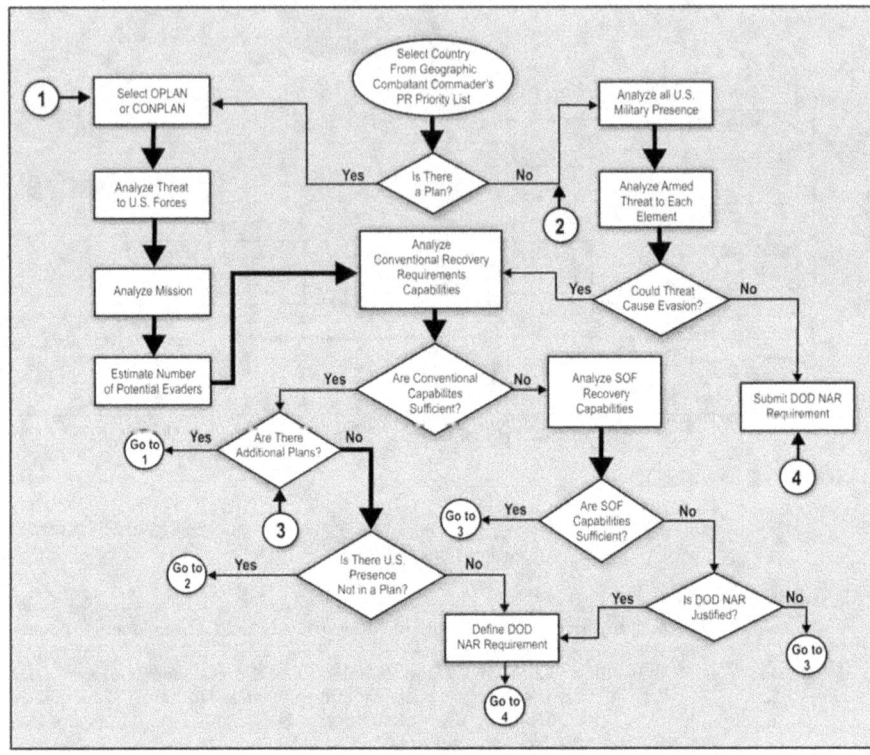

Figure D-3. PR Planning Matrix

COA ANALYSIS, SELECTION, AND RECOMMENDATION

D-12. The tactical planners will evaluate all potential COAs for adequacy, suitability, time constraints, and risk versus gain. Tactical planners will rank-order multiple COAs and present the recommendations for JFSOCC approval. Given JFSOCC, and when necessary, JFACC or JFC approval, the JSRC will transmit a SARREQ. The JFSOCC will direct the transmission of an execute order (EXORD) to the ground tactical elements through their respective C2 nodes.

Appendix E

Personnel Recovery Planner Checklists and Voice Message Templates

This appendix provides example checklists for PR planners (Figures E-1 through E-10, pages E-1 through E-17). It also provides example templates for the SARIR and SARSIT (Figures E-11 and E-12, pages E-18 through E-21). The reporting element completes the SARIR or SARSIT voice message templates and submits them to the JSRC or appropriate component RCC.

STAFF PR PLANNING CHECKLIST
PERSONNEL SECTION (S-1)
____ Assist in preparation of Commander's EPA Guidance, Annex W (Evasion Plan of Action OPLAN) to OPORD XXXXX.
____ Coordinate with the SF PR coordinator regarding personnel accountability and reporting.
____ Coordinate with the SF PR coordinator and SOC RCC regarding casualty affairs for repatriating remains of recovered MIA and KIA personnel.
____ Ensure the S-1 is an information addressee on message traffic regarding downed, missing, or rescued personnel.
____ Coordinate with the joint force legal office to ensure necessary actions are taken to meet all statuary requirements with respect to Service reporting procedures and boards of inquiry.
____ Coordinate with the SOC RCC regarding family support for isolated personnel.
____ Support isolated personnel during the repatriation process (pay, promotions, and personnel actions).
____ Ensure the public affairs officer (PAO) provides information to the news media only after considering:
____ Security requirements.
____ Welfare of returned personnel and their families.
____ The safety and interests of other personnel who may still be detained.
____ Coordinate with the joint legal office for the preparation of legal annexes and appendixes.
____ Ensure all offices involved in PR operations are using standard forms and message formats.
INTELLIGENCE SECTION (S-2)
____ Prepare Intelligence Annex to the Commander's EPA Guidance, and brief at staff mission brief.
____ Provide the SFODA with intelligence estimates and updates, as required, on:
____ Enemy policy, practices, and intentions.
____ Location, strength, capabilities, and activities of enemy units.
____ Enemy early warning systems.
____ Enemy air defense units.
____ Current enemy air, ground, and naval OB.

Figure E-1. Staff PR Planning Checklist

____ Attitude of the populace.
____ JPRSPs, SAFEs, SAIDs, DARs, contact points, ID codes, and designated HLZs.
____ Terrain information and analysis.
____ Coordinate with available intelligence assets to provide the SFODA with timely, objective, and cogent requested military intelligence for PR planning.
____ Ensure the SF PR coordinator conducted preincident coordination for submitting ISOPREP data and EPAs to SOC RCC.
____ Ensure intelligence requirements for PR planning and operations are forwarded to JSOTF J-2.
____ Ensure the command submits a prioritized request to JSOTF for all-source evaluation of evasion areas where ground reconnaissance is not feasible.
____ Ensure copies of SAFE area studies and/or E&R country studies have been requested from JSOTF J-2.
____ Forward to the JSOTF internally developed intelligence requirements relating to PR.
____ Ensure imagery of potential SAFEs are available. If not, submit a request to JSOTF J-2.
____ Ensure information about DARs and SAFEs and their contact points are disseminated to the high-risk-of-capture personnel and aircrews.
____ Ensure the OPCEN and SFODAs receive updates on enemy OB and situation for integration in EPA planning and PR operations.
____ Ensure SERE contingency guides (if produced) have been distributed to the SFODAs.
____ Ensure high-risk-of-capture and potentially isolated personnel are briefed on specific regional aspects that will assist them in their SERE efforts.
____ Coordinate with the joint force PSYOP officer on ways to favorably influence the local population regarding CSAR efforts.
____ Ensure the PR coordinator is included in weather and astronomical data distribution.
____ Ensure all personnel are using maps with identical grid systems and reference points.
____ Coordinate with the SF PR coordinator and OPCEN to ensure recovered personnel are debriefed.
OPERATIONS SECTION (S-3)
____ Ensure a designated SF PR coordinator and a PR cell are established within the OPCEN to manage the command PR program.
____ PR cell must be capable of supporting 24-hour operations.
____ Ensure the SFOB or FOB has an established, written, and disseminated PR program that incorporates conventional and UAR capabilities.
____ Ensure PR doctrine, procedures, and capabilities have been incorporated into all operations.
____ Develop Commander's EPA Guidance Annex.
____ Ensure the command has developed and coordinated PR tactics, techniques, procedures, publications, and equipment requirements.
____ Determine who has launch authority, border-crossing authority, and authority to suspend searches and rescue efforts for isolated personnel.
____ Ensure the responsibilities and procedures for the command SF PR coordinator and the PR cell have been delineated.

Figure E-1. Staff PR Planning Checklist (Continued)

____ Ensure PR planning and operations have been integrated and rehearsed in previous training exercises.
____ Ensure the potential involvement of allied forces in the command PR program has been addressed.
____ Ensure all elements have been briefed on the command PR planning and operations.
____ Ensure all applicable regulations and references are on hand in the OPCEN.
____ Ensure UARM requirements have been established and forwarded to the JSOTF staff and UARCC staff for coordination.
____ Ensure PR issues have been integrated into PSYOP and CA operations.
____ Ensure facilities and equipment for the SF PR coordinator and PR cell are available.
____ Ensure the SOC RCC and UARCC are informed of all current and future operations in the AOR, and provide them applicable map overlays and graphics.
____ Ensure the S-3 or OPCEN is on the daily distribution list for the ATO.
____ Ensure CSAR SPINS are included in the Commander's EPA Guidance.
____ Ensure all personnel are familiar with ATO word, letter, and number of the day, and RAS. This knowledge will facilitate the authentication of all isolated personnel.
____ Ensure the SF PR coordinator and OPCEN monitors ongoing SAR or CSAR operations to avoid conflict or unnecessary complication.
SF PR COORDINATOR
____ Ensure the OPCEN is passing information to the SOC RCC regarding personnel who cannot be recovered by command assets because of logistical, operational, or political constraints.
____ Ensure the command has a published standard designating which personnel are required to receive Levels A, B, and C SERE training.
____ Ensure high-risk-of-capture and isolation personnel receive appropriate Geneva Convention, Code of Conduct, and SERE training.
____ Ensure high-risk-of-capture and isolation personnel are briefed on specific regional aspects that will assist them in their SERE efforts.
____ Ensure high-risk-of-capture personnel have trained using their E&R equipment.
____ Ensure high-risk-of-capture personnel receive training on recovery equipment and techniques of other components.
____ Serve as blood chit program manager for the command, issuing and tracking all blood chits.
____ If blood chits and other equipment and devices to assist personnel in SERE efforts have been issued, ensure there are additional items in stock to replace used, lost, or destroyed equipment.
____ Ensure blood chit accountability records are submitted to the JPRA.
____ Ensure ISOPREPs (DD Form 1833) are prepared and forwarded to the SOC RCC and UARCC, when required.
____ Ensure preincident coordination is made for submitting ISOPREP data and EPAs to SOC RCC and UARCC.
____ Ensure PR planning and operations have been integrated and rehearsed in previous training exercises.
____ Following the notification of an isolating incident, a "flash" priority SARIR should be issued. OPCEN PR cell should issue the SARIR. Ensure this happens. Do not wait until you have all the information required to complete every field on the SARIR. Report the information you have, and issue updates as situation develops.

Figure E-1. Staff PR Planning Checklist (Continued)

___ Ensure training for PR cell personnel is conducted during periods when SAR and CSAR missions are not being conducted.

___ For effective support to PR, SFOB or FOB must provide EEIs, such as aircraft type, recovery activation signals, EPAs, flight route and altitude, and specific SAFEs and E&R areas briefed to the aircrew to the SOC RCC, when necessary.

___ Interface with OCE, as necessary, to support UAR.

___ Maintain connectivity with the UARCC.

LOGISTICS SECTION (S-4)

___ Coordinate with isolated SFODAs for construction of resupply and E&R bundles and caches.

___ Ensure facilities and equipment for PR cell are available.

___ If blood chits and other equipment and devices to assist personnel in SERE efforts have been issued, ensure there are additional items in stock to replace used, lost, or destroyed equipment.

___ Ensure units request additional or new E&R assistance devices when a utility has been discovered.

___ Ensure transportation requirements for repatriated or recovered personnel have been coordinated.

FUTURE OPERATIONS AND PLANS (S-5)

___ Coordinate with the SF PR coordinator and SOC RCC to ensure PR is addressed and included in all OPLANs and OPORDs.

___ Coordinate with OCE and UARCC, as necessary, for UAR operations planning.

COMMUNICATIONS AND COMPUTER SYSTEMS (C-E OR SIGNAL DETACHMENT)

___ Provide the PR cell communications support as outlined below:

___ Voice capability (secure and nonsecure telephones with worldwide access or radio (SATCOM or UHF).

___ Data capability (message, SIPRNET, NIPRNET, and so on).

___ Fax (secure and nonsecure) capability.

___ AN/PRC-112 and CSEL PLS codes.

___ Provide PR support by providing the PR cell or OPCEN with:

___ Networks.

___ Frequencies.

___ Code words.

___ Call signs.

___ Ensure the PR cell communications network is established.

___ Coordinate with the SF PR coordinator to ensure adequate primary and secondary communications links have been established with the SOC RCC and UARCC.

___ Ensure Channels A and B frequencies for the AN/PRC-112 have been confirmed and are programmed into all assigned radios.

___ Ensure all centers involved in PR operations are using standard forms and message formats.

Figure E-1. Staff PR Planning Checklist (Continued)

____ Ensure frequencies, call signs, and communications procedures for the command PR operations have been coordinated and issued to all.
STAFF JUDGE ADVOCATE (SJA)
____ Provide legal information to the command in reference to—
____ Operational law.
____ Law of armed conflict.
____ ROE.
____ Law of the sea.
____ Airspace law.
____ SOFAs.
____ Assist the S-3 with the preparation of ROE appendix to all OPLANs and OPORDs.
____ Assist the S-1 with the preparation of Appendix ___ (Legal) to SFOB OPLAN.
PUBLIC AFFAIRS (PAO)
____ Prepare Annex ___ (Public Affairs) to SFOB OPLAN.
____ Tailor PA activity to support missions across the entire range of military operations.
____ Establish PR-related information goals according to command guidance.
____ Coordinate with the CA and PSYOP staffs to ensure efforts are coordinated.
____ Coordinate with the SF PR coordinator for any public appearances by returnees.
____ Ensure PR-related information for release to the news media has been approved by the JPRA. This information is coordinated through the SF PR coordinator. Information can be released only after considering—
____ Security requirements (OPSEC).
____ Welfare of returned personnel and their families.
____ The safety and interests of other personnel who may still be detained.
SURGEON
____ Prepare Annex ___ (Medical Services) to the SFOB OPLAN or FOB OPORD.
____ Coordinate with the SF PR coordinator regarding PR support concerning—
____ Medical evacuation.
____ Health service support for PR operations.
____ Phases I to V care for recovered personnel.

Figure E-1. Staff PR Planning Checklist (Continued)

INTELLIGENCE SECTION (S-2) PREOPERATIONS CHECKLIST
NOTE: S-2 sections should accomplish as much of this checklist as they can before deployment. Some items must be completed in-theater. In addition, if support personnel are available, their duties may include some or all of the communications and administrative items on this checklist.
____ Obtain more than adequate numbers of all applicable MC&G products. This includes wall charts covering the entire AO and larger scale displays of applicable SAFEs, major population centers, and target-rich areas within the AO. If possible, also acquire electronic format copies and the capability to print and display them. Check which datum and map editions are currently used in the theater to ensure compatibility.
____ Review intelligence requirements for PR. Check their status and upgrade priority, as necessary. As you review old requirements and develop new ones, keep in mind the JSRC may require intelligence support for SERE guidance development, locating isolated personnel, and the allocation and planning of recovery missions.
____ Ensure you have adequate copies of all applicable SAIDs and E&R country and regional studies. Soft copy is available on SIPRNET at http://delphi-s.dia.smil.mil/intel/oicc/twj/twj2_home.htm. Hard copy is available via online request of DIA Support for the Analyst's File Environment. If you do not have access to the DIA site, submit all requests via your normal S-2 channels. If you have operational support requirements and must have the hard copy products within 14 days, contact the DIA/OICC at commercial (202) 231-4945/4944 or DSN 428-4945/4944 Forward any requirements for production of these documents to DIA through theater or command channels.

____ Check the JPRA INTELINK-S (SIPRNET) website at http://jpra.jfcom.smil.mil to ensure that you have adequate copies of all applicable SERE contingency guides, bulletins, newsletters, and update messages. The joint personnel recovery support packages are also accessible via the JPRA SIPRNET web site. If you have questions, contact the analysis and production division of the JPRA Intelligence Support Office. Contact information for all JPRA offices is on the web site at http://jpra.jfcom.smil.mil (SIPRNET).
 	
 Mailing address is:
  HQ JPRA/ISOA
  BLDG 358
  10244 BURBECK ROAD
  FT BELVOIR VA 22060-5805
 	
If all else fails, please call DSN 654-2208 or commercial (703) 704-2208. Requests for new production must be command-validated.

____ Determine whether you require training in PR planning. If so, forward requests for training through the unit SF PR coordinator.
____ Coordinate procedures for debriefing recovered evaders or captives with JTF or theater J-2 and JPRA or DPMO.
____ Review available imagery. Determine if imagery of SAFEs, major population centers, and other target-rich areas is available, current, and of adequate scale and resolution. Submit new requirements or changes to current requirements through normal command channels.
____ Study any other available information on AO relevant to PR. Provide intelligence support to preparatory recovery planning and development.

Figure E-2. Preoperations Checklist for Intelligence Section

_____ Coordinate with the JSOTF J-2 communications regarding message dissemination. Check address lists and message shells for SPINS, ATOs, CONPLANs, communications plans, SOPs, theater SARIRs, and any changes to these documents. Ensure current addresses for all appropriate units are listed.
_____ Ensure connectivity with the following will be available in the theater:
— J-2, JIC, JAC, JISE, and JSOTF, as applicable.
— Weather.
— Theater and JTF representatives for NSA, NRO, NIMA, DIA, JPRA, and other appropriate national agencies.
— Intelligence databases and communications systems.
— SOC RCC intelligence sections.
_____ Make contingency plans for when the primary means of communication fails. Establish contact with above units and agencies, and test everything upon arrival in the theater.
_____ SFOB or FOB S-2 requires ready access to—
_____ A. Secure telephone.
_____ B. Desktop access to INTELINK-S (SIPRNET).
_____ C. Appropriate COMSEC tapes, keys, and so on.
_____ D. Word processing, message development, and briefing capabilities.
_____ E. Unsecure and secure fax capability.
_____ F. Unsecure and secure data capability.
_____ G. DOD national intelligence community support to PR documents:
— JP 2-0, *Unified Action Armed Forces (UNAAF).*
— JP 2-01, *Appendix C to Joint Intelligence Support to Military Operations.*
— JP 2-02, *National Intelligence Support to Military Operations.*
— JP 3-50.2, *Doctrine of Joint Combat Search and Rescue.*
— JP 3-50.21, *Joint Doctrine for Combat Search and Rescue.*
— JP 3-50.3, *Joint Doctrine for Evasion and Recovery.*
— SPINS.
— ATO.
— CONPLANs.
— Communications plans.
— SOPs.
_____ I. CNN and local radio and TV broadcasts.
_____ H. Secure storage (GSA-approved safe or vault, cleared to TS and SCI).

Figure E-2. Preoperations Checklist for Intelligence Section (Continued)

___ Start a contact list of all offices and personnel you may need to reach quickly. Include a recall plan for section personnel. Update it continually.
___ Review all applicable instructions on release of classified information to allied personnel. Ensure the responsible security office educates the SFOB or FOB staff on these procedures. Coordinate with allied units, as appropriate.
___ Upon arrival, check security and soundproofing of work area.
___ Safeguard classified materials. Establish destruction procedures.
___ Establish communications. Send an on-station report to home station and appropriate units and agencies. Include your phone numbers, call signs, GENSER message address, location, POCs, and so on. Request a reply to test all communications channels.
___ Locate other local intelligence organizations, particularly those required to support you. Get latest intelligence and local area situation. Review their chart and imagery stocks, briefing schedule, and connectivity. Establish a positive working relationship.
___ Determine local procedures for obtaining equipment and supplies.
___ Post situation, threat, and OB displays. If electronic displays are used, ensure manual backups are readily available. Displays and overlays should include: friendly and enemy NOB, GOB, AOB, and MOB; borders; FEBA and FLOT, if applicable; evader locations; prisoner locations; NBC locations; no, low-, medium-, and high-threat areas; spider routes and points; SOF corridors; SARDOTs and SAR bull's-eyes; "current as of" DTG; classification; legend; and opaque cover for quick sanitation. Isolated personnel location displays should include the source and DTG of the information.
___ Coordinate shift change times and situation briefing schedule for the OPCEN and Commander's Updates. Recommend one PR briefing per shift (more if necessary).
___ Start an intelligence journal. Include taskings, significant events, and incoming and outgoing communications (telephone calls, GENSER messages, and computer messages). The incoming and outgoing communications will include INTREPs, INTSUMs, DISUMs, SARIRs, SARSITs, RFIs, responses to RFIs, and so on.
___ Set up chart and imagery stocks and distribution procedures.
___ Practice the steps of the Intelligence Incident Checklist.

Figure E-2. Preoperations Checklist for Intelligence Section (Continued)

PR INTELLIGENCE SECTION INCIDENT CHECKLIST
___ Upon notification of isolated personnel, plot the location on the primary situation display chart with the DTG and source of the report.
___ Immediately call OPCEN and SF PR coordinator. Provide the following information: time of incident, location, call sign, and primary SAR frequency.
___ If notification came in through intelligence channels, ensure all OPCEN or SUPCEN personnel are immediately aware of the event.
___ Assess threat situation near and en route to the isolated personnel. Brief the SF PR coordinator, OPCEN director, and others, as necessary.
___ Ensure specific information regarding the isolated personnel is disseminated to the SOC RCC, JSOTF J-2, and others, as directed.

Figure E-3. PR Intelligence Section Incident Checklist

____ Draft RFIs, and issue through established theater channels.
____ Notify appropriate personnel of responses to RFIs and other relevant intelligence updates as they come in. This will include the OPCEN, PR cell, and others, as necessary. Plot changes on display charts.
____ Assist OPCEN planners and PR cell in determining ingress and egress routes, and establishing spider routes and RFAs around the isolated personnel.
____ Identify intelligence assets available to support recovery.
____ Recommend actions to degrade or eliminate threat, if possible.
____ As necessary, provide support to intelligence sections of PR-tasked units.
____ Checklist complete.

Figure E-3. PR Intelligence Section Incident Checklist (Continued)

PR CELL INTELLIGENCE SECTION DAILY OPERATIONS CHECKLIST
____ Attend shift changeover briefing, and review PR cell and OPCEN event logs.
____ Review OPCEN S-2 intelligence journal and read messages from previous shift.
____ Review current ATO, changes to SPINS, and U.S., allied, or coalition missions currently in the planning stages. The SF PR coordinator should have access to this information. Discuss potential threats to these missions with OPCEN planners, so they can determine the appropriate readiness posture of recovery forces.
____ Obtain the daily JIC or JISE intel SITREP.
____ Get 12- and 24-hour weather briefings.
____ Complete unfinished tasks from previous shift; for example, preparing an RFI or plotting a new location for a SAM site.
____ Review open-incident folders and mission folders with detachment LNOs. Check for accuracy and consistency with logs and displays.
____ Review threats to open missions. Ensure SF PR coordinator and OPCEN are aware of the current threat situations for all missions.
____ Brief each shift on current threat situation and probable enemy COA with respect to isolated personnel and PR missions. You may also include significant military and political events; areas of major engagement; weather; local area situation; THREATCON, DEFCON, and MOPP levels; indications of impending attack; indications of NBC activity; and any other special-interest items deemed appropriate by the OPCEN.
____ With S-2 personnel, determine any new EEIs. Pass as RFIs IAW organizational procedures.
____ Check incoming message traffic and other sources of information. Look for information that may affect recovery forces or potential isolated personnel. Maintain threat displays and the PR intelligence journal.
____ Immediately brief critical information to appropriate OPCEN personnel.
____ Maintain awareness of available PR and intelligence assets.
____ Check communications with theater units, JSOTF, SOC RCC, UARCC, and JIC or JISE at least weekly.
____ Prepare shift changeover briefing.
____ Checklist complete.

Figure E-4. PR Cell Intelligence Section Daily Operations Checklist (Continued)

PR CELL INTELLIGENCE SECTION	
SHIFT CHANGE BRIEFING CHECKLIST	
____ Current as of time.	
____ Current status of each isolated individual.	
____ A. Missions.	
____ Completed.	
____ Ongoing.	
____ Planned.	
____ B. Recent incidents.	
____ Local area threat situation and significant events (for example, THREATCON, MOPP, and DEFCON levels).	
____ Significant political events (surrenders, treaties, declarations of war, policy changes, and so on).	
____ Briefed any changes to:	
____ A. Orders of battle.	
____ B. Tactics.	
____ C. Readiness posture.	
____ D. Areas of engagement.	
____ Other significant military activity.	
____ NBC attacks or indications of impending attacks.	
____ Probable course of enemy action.	
____ Results of debriefings.	
____ Significant terrorist activity.	
____ Reviewed the following administrative items. Referenced intelligence journal, as appropriate.	
____ A. Messages, in and out of the PR cell or OPCEN.	
____ B. Communications problems.	
____ C. Unfinished tasks (RFIs, OB plots, and so on).	
____ D. Upcoming events (briefings and so on).	
____ E. Significant events.	
____ F. Status of supplies.	

Figure E-5. PR Cell Intelligence Section Shift Change Briefing Checklist

FM 3-05.231

PR CELL OPERATIONS CHECKLIST
____ Ensure that a coordinated SFOB or FOB E&R program is developed and that all assigned forces are prepared to execute their responsibilities.
____ Establish UARM requirements, and forward them to the JSOTF for coordination.
____ Maintain direct and continuous liaison with designated recovery assets and component command E&R offices of primary responsibility.
____ Ensure EPAs are developed for all operations.
____ Identify the standardized E&R operations procedures involving conventional forces, SF, and UARMs, to include contact and authentication procedures.
____ Ensure all E&R equipment and TTPs are compatible across the force and are disseminated to potential users.
____ Establish procedures to monitor mission progress and status of recovery assets.
____ Establish procedures to locate and communicate with personnel evading in denied or hostile areas.
____ Determine evasion aid requirements.
____ Recommend to the SFOB or FOB commander E&R aids necessary to support component PR requirements.
____ Ensure E&R scenarios during field exercises are realistic and adequate and reflect the theater environment and operating conditions.
____ Participate in the development of E&R tactics, including exercise planning.
____ Maintain records containing all available data on assigned personnel isolated in hostile territory conducting E&R. When no longer needed by the PR cell, forward these records to the JPRA for permanent archiving. Do not destroy any records or case files relating to missing, captured, or recovered personnel.
____ Ensure available data, including all-source intelligence, on the evasion environment in the theater is collected, maintained, and disseminated to mission planners.
____ Forward all information concerning sightings of missing, evading, or captured personnel to SOC RCC and JSOTF.
____ Provide selected personnel for specialized briefings or training when so directed.
____ Obtain current JCSAR SOPs.
____ Obtain current CSAR COMM plans.
____ Obtain established reporting requirements for the SOC RCC and JSRC.
____ Assist in the development of CSAR appendixes to OPLANs, OPLANs in concept format, and OPORDs. Ensure CSAR appendixes are linked to related appendixes for casualty affairs, medical, repatriation, and mortuary affairs.
____ Coordinate and deconflict all unit E&R plans. Review them for supportability, and advise the OPCEN director.
____ Conduct or provide on-the-job informal training for PR cell augmentation personnel, as required.
____ Organize and conduct PR mission training exercises for the command.
____ Develop personnel augmentation requirements.
____ Develop a plan to transition from peacetime to combat operations.
____ Establish additional communications support requirements.

Figure E-6. PR Cell Operations Checklist

___	Ensure the entire PR communications support system is fully operational and has been tested with the SOC RCC and/or other supporting elements. Ensure phone numbers, fax numbers, frequencies, e-mail addresses, and so on have been exchanged with appropriate offices.
___	Establish dedicated intelligence support requirements.
___	Obtain current rules of engagement approved by the joint force.
___	Obtain and disseminate ATOs and CSAR SPINS to be followed by all high-risk combatants.
___	Alert appropriate elements of the known or assumed locations of isolated personnel.
___	Keep SOC RCC and UARCC informed on the status of ongoing personnel recovery operations.
___	Maintain a database and file on each isolated individual until recovery is complete. Forward the database and all files to the JPRA. This is done once the recovery mission is complete and the SFOB or FOB no longer has a requirement to maintain the files. The files should not be destroyed.
___	Obtain current theater repatriation plan, and assist (as required) in executing repatriation plans to return recovered personnel to their units or family.
___	Determine who has launch or border-crossing authority and authority to suspend searches and rescue efforts for isolated personnel.
___	Monitor all CSAR incidents within JSOA.
___	Ensure responsibilities or procedures for the command PR coordinator and PR cell are established.
___	Ensure entire command is briefed on the command PR planning and operations.
___	Ensure any SF elements assigned as recovery forces are identified and made aware of their assigned or potential PR taskings.

Figure E-6. PR Cell Operations Checklist (Continued)

PR COORDINATOR AND PR CELL SHIFT CHANGEOVER CHECKLIST
___ Brief the following items to the oncoming shift supervisor:
___ A. Location of:
___ SFOB or FOB commander.
___ OPCEN director.
___ Designated SF PR assets (number and status).
___ B. The current intelligence situation.
___ C. Incidents or missions in progress:
___ 1) Incidents being worked.
___ a. Location of incident.
___ b. Information required to complete mission planning.
___ c. Open checklist items.

Figure E-7. PR Coordinator and PR Cell Shift Changeover Checklist

FM 3-05.231

_____ 2) Missions assigned to recovery forces:	
_____ a. PR mission commander.	
_____ b. Tasked PR recovery unit.	
_____ c. ETD or ATD of PR forces.	
_____ d. Outstanding items required to launch PR forces.	
_____ D. Incidents and missions closed during previous shift.	
_____ E. Communications equipment status (as applicable):	
_____ 1) Phones (STU-III, and so on)	
_____ 2) Computers (SIPRNET and NIPRNET)	
_____ 3) Radios (tactical, CSAR, SATCOM, internal handheld, and so on)	
_____ 4) Field phones within OPCEN or ISOFAC area	
_____ 5) Nonsecure and secure faxes	
_____ 6) SARSAT	
_____ 7) PRMS (if operational)	
_____ F. Messages received and sent during the shift:	
_____ 1) Mission reports to be drafted and sent to HQ.	
_____ 2) Mission reports sent to higher HQ.	
_____ 3) Messages received during the shift.	
_____ 4) Message traffic awaiting pickup.	
_____ 5) Brief special instructions:	
_____ a. Messages for specific individuals.	
_____ b. New read items that change procedures.	
_____ Ensure the PR cell area is clean and neat.	
_____ Ensure supplies are located at each workstation.	
_____ Log "SHIFT CHANGEOVER BRIEFING COMPLETE" on PR cell event log.	
_____ Checklist complete.	

Figure E-7. PR Coordinator and PR Cell Shift Changeover Checklist (Continued)

PR COORDINATOR AND PR CELL CHECKLIST—DAY SHIFT
_____ Review the following before the shift changeover briefing:
_____ A. Commander's information file (CIF).
_____ B. Event logs since the end of last shift.
_____ Attend shift changeover briefing:
_____ A. Review the OPCEN event log.
_____ B. Inventory COMSEC.
_____ C. Rekey secure UHF and STU-III (as required).
_____ D. Change or destroy COMSEC, as required.
_____ E. Check availability of current authenticators.
_____ Review current ATO SPINS:
_____ A. Disseminate changes.
_____ B. Post call sign and radio frequency.
_____ Conduct communications checks on radios and phone lines.
_____ Obtain WX briefing, and update the weather board.
_____ Obtain intelligence briefing for current day's activity.
_____ Review OPCEN wall displays:
_____ A. Assets board.
_____ B. Incident and mission boards.
_____ C. SFODA tracking boards (LNOs).
_____ Review PR open incident and PR mission folders, and determine follow-up actions. Review category and priority to ensure:
_____ A. Number and condition of known personnel are current and accurate.
_____ B. Coordinates of LKP in folder, on plotting chart, and on mission status board are current and accurate.
_____ C. Current threat information is posted.
_____ Check with SIGCEN for any information on radio contact with isolated personnel.
_____ Brief all centers on any preplanned or expected rescue activity for the day.
_____ Brief the commander, as necessary, on missions in progress and/or opened.
_____ Coordinate with SIGCEN to monitor SAR frequencies for evader transmissions.
_____ During periods of minimal activity, place recovery assets on relaxed alert. Inform them when activity resumes.
_____ Review mission folders, if applicable.

Figure E-8. PR Coordinator and PR Cell Checklist—Day Shift

_____ Review day's activities (open or closed, incident or mission), and prepare summary of day's activity, required reports, and so on.
_____ Prepare for shift changeover.
_____ At midnight ZULU time:
_____ A. Close or open daily PR cell event log.
_____ B. Update crypto for radios, as necessary.
_____ C. Destroy classified waste.
_____ Checklist complete.

Figure E-8. PR Coordinator and PR Cell Checklist—Day Shift (Continued)

PR COORDINATOR AND PR CELL CHECKLIST—SWING SHIFT
_____ Review the following before shift changeover briefing:
_____ A. CIF.
_____ B. Event logs since the end of last shift.
_____ Attend shift changeover briefing:
_____ A. Review the OPCEN event log.
_____ B. Inventory COMSEC.
_____ C. Rekey secure UHF and STU-III, as required.
_____ D. Change or destroy COMSEC, as required.
_____ E. Check availability of current authenticators.
_____ Review current ATO SPINS:
_____ A. Disseminate any changes.
_____ B. Post call sign and radio frequency.
_____ Conduct communications checks on radios and phone lines.
_____ Obtain WX briefing, and update the weather board.
_____ Obtain intelligence briefing for swing shift activity.
_____ Review OPCEN wall displays:
_____ A. Assets board.
_____ B. Incident and mission boards.
_____ C. SFODA tracking boards (LNOs).
_____ Review PR open incident and PR mission folders, and determine follow-up actions. Review category and priority to ensure:

Figure E-9. PR Coordinator and PR Cell Checklist—Swing Shift

____ A. Number and condition of known personnel are current and accurate.
____ B. Coordinates of LKP in folder, on plotting chart, and on mission status board are current and accurate.
____ C. Current threat information is posted.
____ Check with SIGCEN for any information on radio contact with isolated personnel.
____ Brief all centers on any preplanned or expected rescue activity for the day.
____ Brief the commander, as necessary, on missions in progress and/or opened.
____ Coordinate with SIGCEN to monitor SAR frequencies for evader transmissions.
____ During periods of minimal activity, place recovery assets on relaxed alert. Inform them when activity resumes.
____ Review mission folders, if applicable.
____ Review day's activity (open or closed and incident or mission), and prepare summary of day's activity, required reports, and so on.
____ At midnight ZULU time:
____ A. Close and open daily PR cell event log.
____ B. Update crypto for radios, as applicable.
____ C. Destroy classified waste.
____ Prepare for shift changeover.
____ Checklist complete.

Figure E-9. PR Coordinator and PR Cell Checklist—Swing Shift (Continued)

PR COORDINATOR AND PR CELL CHECKLIST—MID SHIFT
____ Review the following before shift changeover briefing:
____ A. CIF.
____ B. Event logs since the end of last shift.
____ Attend shift changeover briefing:
____ A. Review the OPCEN event log.
____ B. Inventory COMSEC.
____ C. Rekey secure UHF and STU-III, as required.
____ D. Change or destroy COMSEC, as required.
____ E. Check availability of current authenticators.
____ Review current ATO SPINS:
____ A. Disseminate any changes.
____ B. Post call sign and radio frequency.

Figure E-10. PR Coordinator and PR Cell Checklist—Mid Shift

_____ Conduct communications checks on radios and phone lines.
_____ Obtain WX briefing, and update the weather board.
_____ Obtain intelligence briefing for swing shift activity.
_____ Review OPCEN wall displays:
_____ A. Assets board.
_____ B. Incident and mission boards.
_____ C. SFODA tracking boards (LNOs).
_____ Review PR open incident and PR mission folders, and determine follow-up actions. Review category and priority to ensure:
_____ A. Number and condition of known personnel are current and accurate.
_____ B. Coordinates of LKP in folder, on plotting chart, and on mission status board are correct.
_____ C. Current threat information is posted.
_____ Check with SIGCEN for any information on radio contact with isolated personnel.
_____ Brief all centers on any preplanned or expected rescue activity for the night or next day.
_____ Brief the commander, as necessary, on missions in progress and/or opened.
_____ Coordinate with SIGCEN to monitor SAR frequencies for evader transmissions.
_____ During periods of minimal activity, place recovery assets on relaxed alert. Inform them when activity resumes.
_____ Review mission folders, if applicable.
_____ A. Obtain WX forecast for time of mission. Update the forecast for time of mission.
_____ B. Obtain intelligence for ingress or egress routes for recovery site.
_____ Review night's activity (open or closed and incident or mission), and prepare and transmit summary of night's activity, required reports (SARSIT), and so on.
_____ Early morning activities:
_____ A. Make sure all status boards are up to date.
_____ B. Confirm availability of recovery assets, WX update, and latest intelligence for any planned missions.
_____ At midnight ZULU time, close and then open daily PR cell event log.

Figure E-10. PR Coordinator and PR Cell Checklist—Mid Shift (Continued)

FM 3-05.231

SARIR VOICE TEMPLATE

SAR INCIDENT REPORT

BIG BOY, THIS IS CROWN.

addressee   originator

CROWN, THIS IS BIG BOY. GO AHEAD, OVER.
addressee answers

THIS IS CROWN.

originator responds

FLASH   IMMEDIATE   PRIORITY   ROUTINE

(Underline and transmit the precedence of this message.)

TOP SECRET   SECRET   CONFIDENTIAL   UNCLASSIFIED

(Underline and transmit the security classification of this message.)

LINE 1 (OR) CALL SIGN_____STOVE PIPE 32_____

(Call sign of disabled or lost aircraft, ship, or other.)

LINE 2 (OR) TYPE_____F4E_____

(Type of disabled or lost aircraft, ship, or other.)

LINE 3 (OR) COLOR WOODLAND CAMOUFLAGE_____

(Color of disabled or lost aircraft, ship, or other.)

LINE 4 (OR) ID _____F00132_____

(Aircraft tail or side number, ship hull number, or other number.)

LINE 5 (OR) LOCATION _____1637N12020E_____

(Location of disabled or lost aircraft, ship, or other using bearing and range, GEOREF, UTM, X-Y, or place name.)

Figure E-11. Example SARIR Voice Message Template Format

LINE 6 (OR) QUALIFIER ____ 120 ____ 60NM _____
(Either ACTUAL or ESTIMATED followed by LAND or SEA to qualify the location.)

LINE 7 (OR) TIME __301230Z_____
(Day-time-zone of incident.)

LINE 8 (OR) CAUSE UNKNOWN _____
(Cause of disabled or lost aircraft, ship, or other.)

LINE 9 (OR) PERSONNEL ____ 2 ACTUAL _____
(Count of personnel on board and qualifier ACTUAL or ESTIMATED.)

LINE 10 (OR) STATUS _____ 1 KILLED IN ACTION 1 EVADING _____
(Enter count of personnel and their status.)

LINE 11 (OR) REQUIRE _____ JSRC ASSISTANCE _____
(Enter JSRC OR COMBINED ASSISTANCE if SAR assistance is required.)

LINE 12 (OR) POINT OF CONTACT_____
(Enter the point of contact and number.)

LINE 13 (OR) NARRATIVE _____
_____
_____
_____

LINE 14 (OR) TIME _____
(Day-hour-minute-zone-month-year, when required to identify the message time of origin.)
LINE 15 (OR) AUTHENTICATION IS _____
(Message authentication in accordance with established procedures.)
OVER.

Figure E-11. Example SARIR Voice Message Template Format (Continued)

```
                           SARSIT VOICE TEMPLATE

          SEARCH AND RESCUE

          SITUATION SUMMARY


_____    RANGER, THIS IS SMOKEY, OVER.
           addressee      originator


_____    SMOKEY, THIS IS RANGER.  GO AHEAD WITH REPORT, OVER
addressee answers, then


THIS IS SMOKEY.    SARSIT AS FOLLOWS
originator responds


     FLASH          IMMEDIATE       PRIORITY      ROUTINE
(Underline and transmit the precedence of this message.)


     TOP SECRET         SECRET         CONFIDENTIAL         UNCLASSIFIED
(Underline and transmit the security classification of this message.)
SAR SITUATION SUMMARY


LINE 1 (OR) MISSION NUMBER _____ SAR 059 _____
(Enter the JSRC SAR mission number.)


LINE 2 (OR) STATUS _____ IN PROGRESS _____
(SAR status: IN PROGRESS, COMPLETED [also for when activity has ceased and will not be resumed], or SUSPENDED if SAR
activity is discontinued and objective was not recovered.)
```

Figure E-12. Example SARSIT Voice Message Template Format

LINE 3 (OR) CALL SIGN _____ STOVE PIPE 32 _____

(Call sign of disabled or lost aircraft, ship, or other.)

LINE 4 (OR) TYPE _____ F4E _____

(Type of disabled or lost aircraft, ship, or other.)

LINE 5 (OR) LOCATION _____ 1637N12020E _____

(Location of SAR incident using bearing and range, GEOREF, lat/long, UTM, X-Y, or place name.)

LINE 6 (OR) PERSONNEL _____ 2 _____

(Number of personnel involved in incident.)

LINE 7 (OR) PERSONNEL STATUS _____ MISSING _____

(Status of personnel involved in incident; for example, recovered.)

LINE 8 (OR) NARRATIVE KING 21 SEARCHING SEA AREA 30 NAUTICAL MILES VICINITY OF LAST REPORTED LOCATION.
_____
_____

LINE 9 (OR) TIME _____

(Day-hour-minute-zone-month-year, when required to identify the message time of origin.)

LINE 10 (OR) AUTHENTICATION IS _____

(Message authentication in accordance with established procedures.)

OVER.

Figure E-12. Example SARSIT Voice Message Template Format (Continued)

Glossary

ABCCC	**airborne battlefield command and control center**—A USAF aircraft equipped with communications, data link, and display equipment; it may be employed as an airborne command post or a communications and intelligence relay facility.
ACC	air component commander
act of mercy	In evasion and recovery operations, assistance rendered to evaders by an individual or elements of the local population who sympathize or empathize with the evaders' cause or plight.
AFDD	Air Force doctrine document
AFI	Air Force instruction
AFR	Air Force regulation
AFSOC	**Air Force Special Operations Command**—The USAF component of a joint force SO component.
AFSOC/CC	Commander, Air Force Special Operations Command
AFSOF	**Air Force special operations forces**—Those Active and Reserve Component Air Force forces designated by the Secretary of Defense that are specifically organized, trained, and equipped to conduct and support special operations. (JP 1-02) Included under AFSOF management and Service proponency are Reserve Component PSYOP units.
AFTTP	Air Force tactics, techniques, and procedures
AGAS	air ground aid section
airborne alert	A state of aircraft readiness wherein combat-equipped aircraft are airborne and ready for immediate action. It is designed to reduce reaction time and to increase the survivability factor.
air defense identification zone	Airspace of defined dimensions within which the ready identification, location, and control of airborne vehicles are required. Commonly referred to as ADIZ.
Air Force special operations detachment	A squadron-size headquarters that could be a composite organization composed of different Air Force special operations assets. The detachment is normally subordinate to an Air Force special operations component, joint special operations task force, or joint task force, depending upon size and duration of the operation. Also known as AFSOD. (JP 1-02)
air operations center	The principal air operations installation from which aircraft and air warning functions of combat air operations are directed, controlled, and executed. It is the senior agency of the Air Force Component Commander from which command and control of air

	operations are coordinated with other components and Services. Also known as AOC. (JP 1-02)
airspace control order	An order implementing the airspace control plan that provides the details of the approved requests for airspace control measures. It is published either as part of the air tasking order or as a separate document. Also known as ACO. (JP 1-02)
airspace control plan	The document approved by the joint force commander that provides specific planning guidance and procedures for the airspace control system for the joint force area of responsibility and/or joint operations area. Also called ACP. (JP 1-02)
ALT	alternate (EPA)
AM	amplitude modulation
AMC	**airborne mission commander**—The AMC serves as an airborne extension of the JSRC and coordinates the CSAR effort between the CSARTF (combat search and rescue task force) and the JSRC by monitoring the status of all combat elements, requesting additional assets when needed, and ensuring the recovery and supporting forces arrive at their designated areas to accomplish the CSAR mission. The AMC may be designated by any level of command and is assigned forces with which to conduct specified CSAR operations. The AMC appoints or relieves, as required, an On-Scene Commander. (JP 3-50.2)
amphibious objective area	A geographical area, delineated in the initiating directive, for purposes of command and control within which is located the objective(s) to be secured by the amphibious task force. This area must be of sufficient size to ensure accomplishment of the amphibious task force's mission and must provide sufficient area for conducting necessary sea, air, and land operations.
amphibious task force	The task organization formed for the purpose of conducting an amphibious operation. The amphibious task force always includes Navy forces and a landing force, with their organic aviation, and may include Military Sealift Command-provided ships and Air Force forces when appropriate.
antiterrorism	Defensive measures used to reduce the vulnerability of individuals and property to terrorist acts, to include limited response and containment by local military forces. See also counterterrorism.
AO	**area of operations**—That portion of an area of war necessary for military operations and for the administration of such operations.
AOB	advanced operational base
AOR	area of responsibility
AR	Army regulation

Army special operations component	The Army component of a joint force special operations component. Also known as ARSOC. (JP 1-02)
ARSOC	Army special operations component
ARSOF	**Army special operations forces**—Those Active and Reserve Component Army forces designated by the Secretary of Defense that are specifically organized, trained, and equipped to conduct and support special operations. (JP 1-02) NOTE: The term "Active Component" has been replaced by "Active Army."
ARSOTF	Army special operations task force
ASAP	as soon as possible
ASOT	advanced special operations techniques
assistance mechanism	Individuals, groups of individuals, or organizations, together with material and/or facilities in position, or that can be placed in position by appropriate U.S. or multinational agencies, to accomplish or support evasion and recovery operations. (JP 1-02)
assisted recovery	The return of an evader to friendly control as the result of assistance from an outside source.
ATD	actual time of departure
ATO	**air tasking order**—A method used to task and disseminate to components, subordinate units, and command and control agencies projected sorties, capabilities, and/or forces to targets and specific missions. Normally provides specific instructions to include call signs, targets, controlling agencies, and so on, as well as general instructions. (JP 1-02)
attached RESCORT	This method allows continuous visual or radar contact of the recovery platform.
attack helicopter	A helicopter specifically designed to employ various weapons to attack and destroy enemy targets.
authentication	1. A security measure designed to protect a communications system against acceptance of a fraudulent transmission or simulation by establishing the validity of a transmission, message, or originator. 2. A means of identifying individuals and verifying their eligibility to receive specific categories of information. 3. Evidence by proper signature or seal that a document is genuine and official. 4. In evasion and recovery operations, the process whereby the identity of an evader is confirmed. (JPs 1-02 and 3-50.2)
auto	automobile
Automatic Secure Voice Communication Network	A worldwide, switched, secure voice network developed to fulfill DOD long-haul, secure voice requirements.
AWACS	Airborne Warning and Control System

blood chit	A small sheet of material depicting an American flag and a statement in several languages to the effect that anyone assisting the bearer to safety will be rewarded. See also evasion aid.
BLS	beach landing site
bn	battalion
bona fides	Good faith. In evasion and recovery operations, the use of verbal and visual communications by individuals who are unknown to one another to establish their authenticity, sincerity, honesty, and truthfulness. (JP 1-02)
BOS	battlefield operating systems
brevity code	A code that provides no security but that has as its sole purpose the shortening of messages rather than the concealment of their content.
C2	command and control
C3I	command, control, communications, and intelligence
C4I	command, control, communications, computers, and intelligence
C4ISR	command, control, communications, computers, intelligence, surveillance, and reconnaissance
CA	**civil affairs**—The activities of a commander that establish, maintain, influence, or exploit relations between military forces and civil authorities, both governmental and nongovernmental, and the civilian populace in a friendly, neutral, or hostile area of operations to facilitate military operations and consolidate operational objectives. Civil affairs may include performance by military forces of activities and functions normally the responsibility of local government. These activities may occur prior to, during, or subsequent to other military actions. They may also occur, if directed, in the absence of other military operations.
cache	In evasion and recovery operations, source of subsistence and supplies, typically containing items such as food, water, medical items, and/or communications equipment, packaged to prevent damage from exposure and hidden in isolated locations by such methods as burial, concealment, and submersion, to support evaders in current or future operations.
CAOC	combined air operations center
CAS	**close air support**—Air action by fixed- and rotary-wing aircraft against hostile targets that are in close proximity to friendly forces and that require detailed integration of each air mission with the fire and movement of those forces.
CAW	carrier air wing
CCRAK	Combined Command for Reconnaissance Activities, Korea
CCT	combat control team

CD	**counterdrug**—Those active measures taken to detect, monitor, and counter the production, trafficking, and use of illegal drugs.
cdr	commander
C-E	communications-electronics
CIA	Central Intelligence Agency
CIF	commander's information file
CJCSI	Chairman, Joint Chiefs of Staff instruction
CJCSM	Chairman, Joint Chiefs of Staff manual
clandestine operation	An operation sponsored or conducted by governmental departments or agencies in such a way as to assure secrecy or concealment. A clandestine operation differs from a covert operation in that emphasis is placed on concealment of the operation rather than on concealment of identity of the sponsor. In special operations, an activity may be both covert and clandestine and may focus equally on operational considerations and intelligence-related activities. (JP 1-02)
CLUES	change of objective report
CMOC	civil-military operations center
co	company
COA	course of action
coalition force	A force composed of military elements of nations that have formed a temporary alliance for some specific purpose.
combat air patrol	An aircraft patrol provided over an objective area, over the force protected, over the critical area of a combat zone, or over an air defense area, for the purpose of intercepting and destroying hostile aircraft before they reach their target.
combatant command	Nontransferable command authority established by Title 10, USC, Section 164, exercised only by commanders of unified or specified combatant commands unless otherwise directed by the President or the Secretary of Defense. Combatant command (command authority) is the authority of a combatant commander to perform those functions of command over assigned forces involving organizing and employing commands and forces, assigning tasks, designating objectives, and giving authoritative direction over all aspects of military operations, joint training, and logistics necessary to accomplish the missions assigned to the command. Combatant command (command authority) should be exercised through the commanders of subordinate organizations; normally this authority is exercised through the Service component commander. Combatant command (command authority) provides full authority to organize and employ commands and forces as the geographic combatant commander considers necessary to accomplish assigned missions. Also known as COCOM.

combat control team	A team of Air Force personnel organized, trained, and equipped to establish and operate navigational or terminal guidance aids, communications, and aircraft control facilities within the objective area of an airborne operation.
combat recovery	The act of retrieving resources while engaging enemy forces.
combat search and rescue mission	The recovery of isolated personnel from contested territory. It involves the tasks of detection, location, identification, authentication, extraction, transportation, and en route medical care of isolated personnel. Also called CSAR mission.
combat search and rescue mission coordinator	The designated person or organization selected to direct and coordinate support for a specific combat search and rescue mission. Also called CSAR mission coordinator.
combat survival	Those measures to be taken by Service personnel when involuntarily separated from friendly forces in combat, including procedures relating to individual survival, evasion, escape, and conduct after capture. (JP 3-50.2)
COMM	communications
command, control, communications, and computer systems	An integrated system of doctrine, procedures, organizational structures, personnel, equipment, facilities, and communications designed to support a commander's exercise of command and control across the range of military operations. Also called C4 systems.
component search and rescue controller	The designated search and rescue representative of a component commander of a joint force who is responsible for coordinating and controlling that component's search and rescue forces. (JP 3-50.2)
COMSEC	communications security—The protection resulting from all measures designed to deny unauthorized persons information of value that might be derived from the possession and study of telecommunications, or to mislead unauthorized persons in their interpretation of the results of such possession and study. Also called COMSEC. (JP 3-50.2)
CONOPS	concept of operations
CONPLAN	operation plan in concept format
contact point	In evasion and recovery operations, a location where an evader can establish contact with friendly forces. (JP 1-02)
contact procedure	Those predesignated actions taken by evaders and recovery forces that permit link-up between the two parties in hostile territory and facilitate the return of evaders to friendly control. (JP 1-02)
contested territory	Includes any landmass or body of water not under friendly control. The airspace over contested territory may, or may not, be under friendly control. Thus, establishing air superiority or air supremacy over the theater of operations may not eliminate contested territory.

CONUS	continental United States
conventional forces	Those forces capable of conducting operations using non-nuclear weapons. Also, those forces not specially trained, equipped, and organized to conduct SO.
conventional recovery operation	Evader recovery operations conducted by conventional forces. (A proposed term identified in JP 3-50.3).
coord	coordination
coordinating authority	A commander or individual assigned responsibility for coordinating specific functions or activities involving forces of two or more Military Departments or two or more forces of the same Service. The commander or individual has the authority to require consultation between the agencies involved, but does not have the authority to compel agreement. In the event that essential agreement cannot be obtained, the matter shall be referred to the appointing authority. Coordinating authority is a consultation relationship, not an authority through which command may be exercised. Coordinating authority is more applicable to planning and similar activities than to operations.
counterinsurgency	Those military, paramilitary, political, economic, psychological, and civic actions taken by a government to defeat insurgency.
counterterrorism	Offensive measures taken to prevent, deter, and respond to terrorism. Also known as CT.
covert operation	An operation that is so planned and executed as to conceal the identity of or permit plausible denial by the sponsor. A covert operation differs from a clandestine operation in that emphasis is placed on concealment of identity of sponsor rather than on concealment of the operation. (JP 1-02) In SO, an activity may be both covert and clandestine.
CPT	captain
crash locator beacon	An automatic emergency radio locator beacon to help searching forces locate a crashed aircraft. (JP 1-02)
crypto	cryptography
CSAR	**combat search and rescue**—A specific task performed by rescue forces to effect the recovery of distressed personnel during war or military operations other than war. (JP 3-50.2)
CSARTF	**combat search and rescue task force**—All forces committed to a specific combat search and rescue operation to search for, locate, identify, and recover isolated personnel during wartime or contingency operations. This includes those elements assigned to provide command and control and protect the rescue vehicle from enemy air or ground attack. (JP 3-50.2)
CSEL	combat survivor evader locator
CW3	chief warrant officer 3

DA	**direct action**—Short-duration strikes and other small-scale offensive actions by special operations forces or special operations-capable units to seize, destroy, capture, recover, or inflict damage on designated personnel or materiel. In the conduct of these operations, special operations forces or special operations-capable units may employ raid, ambush, or direct assault tactics; emplace mines and other munitions; conduct standoff attacks by fire from air, ground, or maritime platforms; provide terminal guidance for precision-guided munitions; conduct independent sabotage; and conduct antiship operations. (JP 1-02)
DAR	designated area for recovery
DCID	Director of Central Intelligence Directive
deception	Those measures designed to mislead the enemy by manipulation, distortion, or falsification of evidence to induce him to react in a manner prejudicial to his interests.
DEFCON	defense readiness condition
DET 101	Detachment 101
detached RESCORT	This method includes route reconnaissance ahead of the recovery platform, trail escort, or proximity escort. Detached RESCORT requires knowledge of routes and planned timing or position radio calls.
DHS	Defense Human Intelligence (HUMINT) Service
DIA	Defense Intelligence Agency
direction finding	A procedure for obtaining bearings of radio frequency emitters by using a highly directional antenna and a display unit on an intercept receiver or ancillary equipment. (JP 3-50.2)
DIRNSA	Director, National Security Agency
DISUM	daily intelligence summary
ditching	Controlled landing of a distressed aircraft on water. (JP 1-02)
DLA	Defense Logistics Agency
DMA	Defense Mapping Agency
DOB	date of birth
DOD	Department of Defense
DODD	Department of Defense directive
DODI	Department of Defense instruction
DOS	Department of State
DPMO	Defense Prisoner of War (POW) Missing Personnel (MP) Office
DS	direct support
DSN	Defense Switched Network

DST	decision support template
DTG	date-time group
duckbutt	An aircraft assigned to perform precautionary search and rescue or combat search and rescue missions, support deployment of single-engine aircraft, or meet other specialized situations. The aircraft can perform a secondary role as navigation aid to passing aircraft. The aircraft is electronically equipped to provide radar tracking, homing, and steering, and gives position and weather reports as required. (JP 3-50.2)
DZ	**drop zone**—A specific area upon which airborne troops, equipment, or supplies are airdropped.
E	east
E&E	**evasion and escape**—The procedures and operations whereby military personnel and other selected individuals are enabled to emerge from enemy-held or hostile areas to areas under friendly control. (JP 3-50.2)
E&R	**evasion and recovery**—The full spectrum of coordinated actions carried out by evaders, recovery forces, and operational recovery planners to effect the successful return of personnel isolated in hostile territory to friendly control.
EALT	earliest anticipated launch time
EDP	evasion decision point
EEI	essential elements of information
emergency locator beacon	A generic term for all radio beacons used for emergency locating purposes.
EPA	**evasion plan of action**—A course of action, developed prior to executing a mission, which is intended to improve a potential evader's chances of successful evasion and recovery by providing recovery forces with an additional source of information that can increase the predictability of the evader's actions and movement. (JP 3-50.2)
ERDZ	emergency resupply drop zone
escort	1. A combatant unit(s) assigned to accompany and protect another force or convoy. 2. Aircraft assigned to protect other aircraft during a mission. 3. An armed guard that accompanies a convoy, a train, prisoner, etc. 4. An armed guard accompanying persons as a mark of honor. (DOD) 5. To convoy. 6. A member of the Armed Forces assigned to accompany, assist, or guide an individual or group, e.g., an escort officer. (JP 3-50.2)
EST	evasion support time line
ETA	estimated time of arrival
ETD	estimated time of departure

evader	Any person in hostile or unfriendly territory who eludes capture. (JP 3-50.2)
evasion	The process whereby individuals who are isolated in hostile or unfriendly territory avoid capture with the goal of successfully returning to areas under friendly control. (JP 3-50.2)
evasion aid	In evasion and recovery operations, any piece of information or equipment designed to assist an individual in evading capture. Evasion aids include, but are not limited to, blood chits, pointee-talkees, evasion charts, barter items, and equipment designed to complement issued survival equipment.
evasion and escape	The process whereby individuals who are isolated in hostile or unfriendly territory avoid capture with the goal of successfully returning to areas under friendly control. (JP 1-02)
evasion and escape intelligence	Processed information prepared to assist personnel to escape if captured by the enemy or to evade capture if lost in enemy-dominated territory.
evasion and escape net	The organization within enemy-held or hostile areas that operates to receive, move, and exfiltrate military personnel or selected individuals to friendly control. Now called a recovery mechanism (RM).
evasion and escape route	A course of travel, preplanned or not, that an escapee or evader uses in an attempt to depart enemy territory to return to friendly lines.
evasion chart	Special map or chart designed as an evasion aid.
evasion decision point	Used in EPA planning to identify any point where the egress or recovery force platform exits the evasion corridor of an SFODA, or travels beyond the boundaries of the EPA overlay of an SFODA. Must be specified during EPA planning. Any location during egress, where the EPA of the SFODA becomes infeasible, thus causing the use of the EPA of the egress platform crew.
EVC	evasion chart
exfil	**exfiltrate or exfiltration**—The removal of personnel or units from areas under enemy control.
EXORD	execute order
fax	facsimile
FEBA	**forward edge of the battle area**—The foremost limits of a series of areas in which ground combat units are deployed, excluding the areas in which the covering or screening forces are operating, designated to coordinate fire support, the positioning of forces, or the maneuver of units.
FEP	final evasion point
FFU	forward friendly unit

fire support area	An appropriate maneuver area assigned to fire support ships from which to deliver gunfire support of an amphibious operation. Also known as FSA.
fire support coordination center	A single location in which centralized communications facilities and personnel incident to the coordination of all forms of fire support are located. Also known as FSCC.
fire support coordination line	A line established by the appropriate ground commander to ensure coordination of fire not under his control but which may affect current tactical operations. The fire support coordination line is used to coordinate fires of air, ground, or sea weapons systems using any type of ammunition against surface targets. The fire support coordination line should follow well-defined terrain features. Also known as FSCL.
FM	frequency modulation; field manual
FOB	forward operational base—In SO, a base usually located in friendly territory or afloat that is established to extend C2 or communications or to provide support for training and tactical operations. Facilities may be established for temporary or longer duration operations and may include an airfield or an unimproved airstrip, an anchorage, or a pier. A forward operations base may be the location of SO component HQ or a smaller unit that is controlled and/or supported by a main operations base.
FOL	forward operating location
foreign internal defense	Participation by civilian and military agencies of a government in any of the action programs taken by another government to free and protect its society from subversion, lawlessness, and insurgency. Also known as FID.
forward air controller	An officer (aviator/pilot) member of the tactical air control party whom, from a forward ground or airborne position, controls aircraft in close air support of ground troops. (JP 3-50.2)
forward area rearming/ refueling point	A temporary facility, organized, equipped, and deployed by an aviation commander, and normally located in the main battle area closer to the AO than the aviation unit's combat service area, to provide fuel and ammunition necessary for the employment of aviation maneuver units in combat. The forward arming and refueling point permits combat aircraft to rapidly refuel and rearm simultaneously. Also known as FARRP.
forward line of own troops	A line that indicates the most forward positions of friendly forces in any kind of military operation at a specific time. The "forward line of own troops" normally identifies the forward location of covering and screening forces.
FRAG order	fragmentary order
G-3	Army or Marine Corps component operations staff officer (Army division or higher staff, Marine Corps brigade or higher staff)

GAS	ground-to-air signal (Also known as GTAS.)
GAT	guidance apportionment targeting
GCCS	Global Command and Control System
GENSER	general service (message)
GEOREF	geographical reference
global positioning system	A system composed of a constellation of satellites and receivers that provides accurate position, time, and movement information to the users of the receivers.
GRAZE	landing zone, drop zone, or pickup zone report (message format)
ground alert	That status in which aircraft on the ground/deck are fully serviced and armed, with combat crews in readiness to take off within a specified short period of time (usually 15 minutes) after receipt of a mission order.
GSA	General Services Administration
guerrilla force	A group of irregular, predominantly indigenous personnel organized along military lines to conduct military and paramilitary operations in enemy-held, hostile, or denied territory.
guerrilla warfare	Military and paramilitary operations conducted in enemy-held or hostile territory by irregular, predominantly indigenous forces. Guerrilla warfare may also be conducted in politically denied areas. Also known as GW.
h	hour
handover/crossover	In evasion and recovery operations, the transfer of evaders between two recovery forces.
helicopter landing site	A designated subdivision of a helicopter landing zone in which a single flight or wave of assault helicopters land to embark or disembark troops and/or cargo.
helicopter team	The combat-equipped troops lifted in one helicopter at one time.
helo	helicopter
Hercules	A medium-range troop and cargo transport designed for airdrop or airland delivery into a combat zone as well as conventional airlift. This aircraft is equipped with four turboprop engines and integral ramp and cargo door. The D model is ski-equipped. The E model has additional fuel capacity for extended range. Designated as C-130. The in-flight tanker configurations are designated as KC-130 and HC-130, which are also used for the aerial rescue mission. The gunship version is designated as AC-130.
HF	high frequency
high-density airspace control zone	Airspace designated in an airspace control plan or airspace control order, in which there is a concentrated employment of numerous and varied weapons and airspace users. A high-density

airspace control zone has defined dimensions, which usually coincide with geographical features or navigational aids. Access to a high-density airspace control zone is normally controlled by the maneuver commander. The maneuver commander can also direct a more restrictive weapons status within the high-density airspace control zone. Also known as HIDACZ.

high-risk-of-capture personnel	U.S. personnel whose position or assignment makes them particularly vulnerable to capture by hostile forces in combat, by terrorists, or by unfriendly governments.
HLZ	**helicopter landing zone**—A specified ground area for landing helicopters to embark or disembark troops and/or cargo. A landing zone may contain one or more landing sites.
HMMWV	high mobility multipurpose wheeled vehicle
HN	host nation
homing	The technique whereby a mobile station directs itself, or is directed, toward a source of primary or reflected energy, or to a specified point.
homing adapter	A device that when used with an aircraft radio receiver, produces aural and/or visual signals that indicate the direction of a transmitting radio station with respect to the heading of the aircraft.
HQ	headquarters
HS	hide site
HU	hole-up site
humanitarian assistance	Programs conducted to relieve or reduce the results of natural or man-made disasters or other endemic conditions such as human pain, disease, hunger, or privation that might present a serious threat to life or that can result in great damage to or loss of property. Humanitarian assistance provided by U.S. forces is limited in scope and duration. The assistance provided is designed to supplement or complement the efforts of the HN civil authorities or agencies that may have the primary responsibility for providing humanitarian assistance.
HUMINT	human intelligence
IAW	in accordance with
ID	identification
IEP	initial evasion point
inc	incident
infil	**infiltrate or infiltration**—1. The movement through or into an area or territory occupied by either friendly or enemy troops or organizations. The movement is made, either by small groups or by individuals, at extended or irregular intervals. When used in connection with the enemy, it infers that contact is avoided. 2. In

	intelligence usage, placing an agent or other person in a target area in hostile territory. Usually involves crossing a frontier or other guarded line. Methods of infiltration are: black (clandestine); gray (through legal crossing point but under false documentation); white (legal).
initial point	An air control point in the vicinity of the LZ from which individual flights of helicopters are directed to their prescribed landing sites. Also known as IP.
inland search and rescue region	The inland areas of the continental United States, except waters under the jurisdiction of the United States. See also search and rescue region.
insurgency	An organized movement aimed at the overthrow of a constituted government through use of subversion and armed conflict.
intel	intelligence
interoperability	1.The ability of systems, units, or forces to provide services to and accept services from other systems, units, or forces, and to use the services so exchanged to enable them to operate effectively together (DOD). 2. The condition achieved among communications-electronics systems or items of communications-electronics equipment when information or services can be exchanged directly and satisfactorily between them and/or their users. The degree of interoperability should be defined when referring to specific cases.
INTREP	intelligence report
INTSUM	intelligence summary
IPB	intelligence preparation of the battlespace
IPG	isolated personnel guidance
IR	infrared; information requirements
ISB	intermediate staging base
ISOFAC	isolation facility
isolated personnel	Military or civilian personnel that have become separated from their unit or organization in an environment requiring them to survive, evade, or escape while awaiting rescue or recovery. (JP 3-50.2)
ISOPREP	**isolated personnel report**—A DOD Form (DD 1833) that contains information designed to facilitate the identification and authentication of an evader by a recovery force.
ITO	integrated tasking order
J-1	personnel directorate
J-2	intelligence directorate
J-3	operations directorate of a joint staff

J-4	logistics directorate of a joint staff
J-5	plans directorate of a joint staff
J-6	command, control, communications, and computer systems directorate of a joint staff
J-32	nonconventional assisted recovery operations cell in JPRA
JAC	joint analysis center
JACK	Joint Advisory Commission, Korea
JAOC	joint air operations center
JCET	joint combined exercise for training
JCO	joint combined operations
JCSAR	joint combat search and rescue
JCSAR forces	Combat forces in the theater of operations that perform one or more roles of the JCSAR mission.
JCS publication	Publications of joint interest applicable to the Services, unified and specified commands, and other authorized agencies prepared under the cognizance of the Joint Staff directorates, or other publications accepted for designation as JCS publications. They are authenticated by the Secretary of JCS "For the Joint Chiefs of Staff" and distributed through Service Channels. JCS publications are approved by the Joint Chiefs of Staff and referred to as "joint publications." (JP 3-50.2)
JFACC	joint force air component commander—The joint force air component commander derives his or her authority from the joint force commander who has authority to exercise operational control, assign missions, direct coordination among subordinate commanders, redirect and organize force to ensure unity of effort in the accomplishment of his or her overall mission. The Joint Force air component commander's responsibilities will be assigned by the joint force commander (normally these would include, but not be limited to, planning, coordination, allocation and tasking based on the joint force commander's apportionment decision). Using the joint force commander's guidance and authority, and in coordination with other service component commanders and other assigned or support commanders, the joint force air component commander will recommend to the joint force commander apportionment of air sorties to various missions or geographic areas. (JP 3-50.2)
JFC	joint force commander—A general term applied to a combatant commander, subunified combatant commander, or joint task force commander authorized to exercise combatant command (command authority), operational control, or tactical control over a joint force. (JP 3-50.2)
JFCC	joint force component commander

JFLCC	**joint force land component commander**—The commander within a unified command, subordinate unified command, or joint task force responsible to the establishing commander for making recommendations on the proper employment of assigned, attached, and/or made available for tasking land forces; planning and coordinating land operations; or accomplishing such operational missions as may be assigned. The joint force land component commander is given the authority necessary to accomplish missions and tasks assigned by the establishing commander. (JP 1-02)
JFMCC	**joint force maritime component commander**—The commander within a unified command, subordinate unified command, or joint task force responsible to the establishing commander for making recommendations on the proper employment of assigned, attached, and/or made available for tasking maritime forces and assets; planning and coordinating maritime operations; or accomplishing such operational missions as may be assigned. The joint force maritime component commander is given the authority necessary to accomplish missions and tasks assigned by the establishing commander. (JP 1-02)
JFSOC	joint force special operations component
JFSOCC	**joint force special operations component commander**—The commander within a unified command, subordinate unified command, or joint task force responsible to the establishing commander for making recommendations on the proper employment of assigned, attached, and/or made available for tasking special operations forces and assets; planning and coordinating special operations; or accomplishing such operational missions as may be assigned. The joint force special operations component commander is given the authority necessary to accomplish missions and tasks assigned by the establishing commander. (JP 1-02)
JIC	joint intelligence center
JISE	joint intelligence support element
JOA	**joint operations area**—An area of land, sea, and airspace, defined by a geographic combatant commander or subordinate unified commander, in which a joint force commander (normally a joint task force commander) conducts military operations to accomplish a specific mission. Joint operations areas are particularly useful when operations are limited in scope and geographic area or when operations are to be conducted on the boundaries between theaters. (JP 1-02)
JOC	**joint operations center**—A jointly manned facility of a joint force commander's headquarters established for planning, monitoring, and guiding the execution of the commander's decisions. (JP 3-50.2)

JOG	joint operations graphic
joint	Connotes activities, operations, organizations, and so on, in which elements of two or more Military Departments participate. (When all Services are not involved, the participating Services shall be identified, for example, Joint Army-Navy.) (JP 3-50.2)
joint combat search and rescue operation	A combat search and rescue operation in support of a component's military operations that has exceeded the combat search and rescue capabilities of that component and requires the efforts of two or more components of the joint force. Normally, the operation is conducted by the joint force commander or a component commander that has been designated by joint force commander tasking. (JP 3-50.2)
joint doctrine	Fundamental principles that guide the employment of forces of two or more Services in coordinated action toward a common objective. It will be promulgated by the Chairman of the Joint Chiefs of Staff, in coordination with the combatant commands, Services, and Joint Staff.
Joint Doctrine Working Party	A forum to include representatives of the Services, combatant commands, and the Joint Staff (represented by the Operational Plans and Joint Force Development Directorate, J-7) which meets semiannually to address and vote on project proposals; discuss key joint doctrinal or operational issues; keep up to date on the status of the joint publication projects and emerging publications; and keep abreast of other initiatives of interest to the members. The Joint Doctrine Working Party meets under the sponsorship of the Director, J-7, Joint Staff. Also called JDWP. (JP 1-02)
joint force	A general term applied to a force composed of significant elements, assigned or attached, of two or more Military Departments, operating under a single joint force commander.
Joint Operation Planning and Execution System	A system that provides the foundation for conventional command and control by national- and combatant command-level commanders and their staff. It is designed to satisfy their information needs in the conduct of joint planning and operations. Joint Operation Planning and Execution System (JOPES) includes joint operation planning policies, procedures, and reporting structures supported by communications and automated data processing systems. JOPES is used to monitor, plan, and execute mobilization, deployment, employment, sustainment, and redeployment activiities associated with joint operations. Also called JOPES. (JP 1-02)
joint operations	A general term to describe military actions conducted by joint forces or by Service forces in relationships (e.g., support, coordinating authority), which, of themselves, do not create joint forces. (JP 1-02)
joint search and rescue center director	The designated representative with overall responsibility for operation of the joint search and rescue center. (JP 1-02)

joint special operations air component commander	The commander within the joint force special operations command responsible for planning and executing join special air operations and for coordinating and deconflicting such operations with conventional nonspecial operations air activities. The joint special operations air component commander normally will be the commander with the preponderance of assets and/or greatest ability to plan, coordinate, allocate, task, control, and support the assigned joint special operations aviation assets. The joint special operations air component commander may be directly subordinate to the joint force special operations component commander or to any nonspecial operations component or joint force commander as directed. Also known as JSOACC. (JP 1-02)
joint tactics, techniques, and procedures	The actions and methods that implement joint doctrine and describe how forces will be employed in joint operations. They are authoritative; as such, joint tactics, techniques, and procedures will be followed except when, in the judgment of the commander, exceptional circumstances dictate otherwise. They will be promulgated by the Chairman of the Joint Chiefs of Staff, in coordination with the combatant commands and Services. Also called JTTP. (JP 1-02)
JP	**joint publication**—A publication containing joint doctrine and/or joint tactics, techniques, and procedures that involves the employment of forces prepared under the cognizance of Joint Staff directorates and applicable to the Military Departments, combatant commands, and other authorized agencies. It is approved by the Chairman of the Joint Chiefs of Staff, in coordination with the combatant commands and Services. Also called JP. (JP 1-02)
JPRA	**Joint Personnel Recovery Agency**—The principal DOD agency for joint personnel recovery support. The JPRA provides joint personnel recovery expertise and assistance to the Office of the Secretary of Defense, the combatant commands, the Chairman of the Joint Chiefs of Staff, the Services, defense agencies, DOD field activities, and other governmental agencies on issues related to CSAR; evasion and recovery (to include NAR and UAR; operational PW/MIA matters; and Code of Conduct training, which includes survival, evasion, resistance, and escape (SERE). (Formerly known as Joint Services SERE Agency.)
JPRSP	Joint Personnel Recovery Support Product
JSCP	Joint Strategic Capabilities Plan
JSOA	**joint special operations area**—A restricted area of land, sea, and airspace assigned by a joint force commander to the commander of joint SOF to conduct special operations activities. The commander of joint SOF may further assign a specific area or sector within the joint special operations area to a subordinate commander for mission execution. The scope and duration of the SOF mission, friendly and hostile situation, and politico-military considerations all influence the number, composition, and

	sequencing of SOF deployed into a joint special operations area. It may be limited in size to accommodate a discrete direct action mission or may be extensive enough to allow a continuing broad range of unconventional warfare operations. (JP 1-02)
JSOAC	joint special operations aviation component
JSOACC	joint special operations air component commander
JSOLE	joint special operations liaison element
JSOTF	**joint special operations task force**—A joint task force composed of special operations units from more than one Service, formed to carry out a specific special operation or prosecute special operations in support of a theater campaign or other operations. The joint special operations task force may have conventional nonspecial operations units assigned or attached to support the conduct of specific missions. (JP 1-02)
JSOTF-NA	Joint Special Operations Task Force—Noble Anvil
JSRC	**joint search and rescue center**—A primary search and rescue facility suitably staffed by supervisory personnel and equipped for planning, coordinating, and executing joint search and rescue and combat search and rescue operations within the geographical area assigned to the joint force. The facility is operated jointly by personnel from two or more Service or functional components or it may have a multinational staff of personnel from two or more allied or coalition nations (multinational search and rescue center). The JSRC should be staffed equitably by trained personnel drawn from each joint force component, including US Coast Guard participation where practical. Formerly called joint rescue coordination center. (JPs 1-02 and 3-50.2)
KIA	killed in action
land search	The search of terrain by earthbound personnel.
lat	latitude
lb	pound
lifeguard submarine	A submarine employed for rescue in an area that cannot be adequately covered by air or surface rescue facilities because of enemy opposition, distance from friendly bases, or other reasons. It is stationed near the objective and sometimes along the route to be flown by the strike aircraft.
LKP	last known position
LNO	liaison officer
load signal	A visual signal displayed in a covert manner to indicate the presence of an individual or object at a given location. (JP 1-02)
long	longitude
low threat	This environment contains threats, whose concentration and capability are such that passive measures enable avoidance.

	Detection is likely to be without consequence even if engaged by weapons. Likely weapons are rocket-propelled grenade, light optically aimed antiaircraft artillery, and weapons up to 14.5 mm.
LT	lieutenant
LZ	landing zone
MACVSOG	Military Assistance Command, Vietnam, Studies and Observation Group
main operations base	In SO, a base established by a JFSOCC or a subordinate SO in friendly territory to provide sustained C2, administration, and logistical support to SO activities in designated areas. Also known as MOB.
MAJ	major
Marine expeditionary unit (special operations capable)	The Marine Corps standard, forward-deployed, sea-based expeditionary organization. The Marine expeditionary unit (MEU) (SO capable) is oriented toward amphibious raids, at night, under limited visibility, while employing emissions control procedures. The Marine expeditionary unit (SO capable) is not a SecDef-designated SO force but, when directed by Defense designated special operations force but, when directed by the President or SecDef and/or the theater commander, may conduct hostage recovery or other SO under in-extremis circumstances when designated SO forces are not available.
maritime environment	The oceans, seas, bays, estuaries, islands, and coastal areas, and the airspace above these, including amphibious objective areas.
maritime search and rescue region	The waters subject to the jurisdiction of the United States; the territories and possessions of the United States (except Canal Zone and the inland area of Alaska) and designated areas of the high seas. (JP 1-02)
MC&G	mapping, charting, and geodesy
MCM	memorandum issued in the name of the Chairman of the Joint Chiefs of Staff
MCOO	modified combined obstacle overlay
MDL	mission decision line
MDMP	military decision-making process
medium threat	This environment contains threats whose concentration and capability are such that active measures must be taken to avoid the threat. Detection is likely to result in engagement by threat weapon systems. Weapons include SAMs, radar-controlled antiaircraft artillery, and air threats lacking look-down/shoot-down. Requires extensive mission planning, evasive maneuvers, avoidance tactics, electronic countermeasures, or defensive threat suppression. May require armed escort. The environment may restrict attack tactics.

METT-TC	mission, enemy, terrain and weather, troops and support available—time available, and civil considerations
MG	major general
MGRS	military grid reference system
MIA	missing in action
MID	military intelligence detachment
minimum-risk route	A temporary corridor of defined dimensions recommended for use by high-speed, fixed-wing aircraft that presents the minimum known hazards to low-flying aircraft transiting the combat zone. Also known as MRR.
mission commander	The commander given TACON over assigned/attached operational and mission support forces, to attain specified mission objectives. The mission commander is responsible for planning, coordinating and executing the operation.
mission type order	1. Order issued to a lower unit that includes the accomplishment of the total mission assigned to the higher headquarters. 2. Order to a unit to perform a mission without specifying how it is to be accomplished. (JP 1-02)
MOA	memorandum of agreement
MOOTW	military operations other than war—Operations that encompass the use of military capabilities across the range of military operations short of war. These military actions can be applied to complement any combination of the other instruments of national power and occur before, during, and after war. (JP 1-02)
MOPP	mission-oriented protective posture
msn	mission
MSR	mission support request
MSS	mission support site
multi-Service doctrine	Fundamental principles that guide the employment of forces of two or more Services in coordinated action toward a common objective. It is ratified by two or more Services, and is promulgated in multi-Service publications that identify the participating Services; e.g., Army-Navy doctrine. (JP 1-02)
NAR	nonconventional assisted recovery—Evader recovery conducted by SOF UW ground and maritime forces and OGAs who are specially trained to develop NAR infrastructure, and interface with or employ indigenous or surrogate personnel. These forces operate in uncertain or hostile areas where CSAR capability is either infeasible, inaccessible, or does not exist to contact, authenticate, support, move, and exfiltrate isolated personnel back to friendly control. NAR forces generally deploy

	into their assigned areas before strike operations and provide the JFC with a coordinated PR capability for as long as the force remains viable.
naval special operations forces	Those Active Navy and Reserve Component Navy forces designated by the SecDef that are specifically organized, trained, and equipped to conduct and support SO. Also called NSW forces or NAVSOF.
naval special warfare	A designated naval warfare specialty that conducts operations in the coastal, riverine, and maritime environments. Naval special warfare emphasizes small, flexible, mobile units operating under, on, and from the sea. These operations are characterized by stealth, speed, and precise, violent application of force. Also called NSW. (JP 1-02)
naval special warfare forces	Those Active and Reserve Component Navy forces designated by the Secretary of Defense that are specifically organized, trained, and equipped to conduct and support special operations. Also called NSW forces or NAVSOF. (JP 1-02)
naval special warfare group	A permanent Navy echelon III major command to which most naval special warfare forces are assigned for some operational and all administrative purposes. It consists of a group headquarters with command and control, communications, and support staff; sea-air-land teams; and sea-air-land delivery vehicle teams. Also known as NSWG. (JP 1-02)
naval special warfare special operations components	The Navy special operations component of a unified or subordinate unified command or joint special operations task force. Also known as NAVSOC. (JP 1-02)
naval special warfare task group	A provisional naval special warfare organization that plans, conducts, and supports special operations in support of fleet commanders and joint force special operations component commanders. Also known as NSWTG. (JP 1-02)
naval special warfare task unit	A provisional subordinate unit of a naval special warfare task group. Also known as NSWTU. (JP 1-02)
naval special warfare unit	A permanent Navy organization forward based to control and support attached naval special warfare forces. Also called NSWU. (JP 1-02)
NAVSOC	naval special operations component
NBC	nuclear, biological, and chemical
NCO	noncommissioned officer
NEO	**noncombatant evacuation operation**—Operations directed by the DOS, the DOD, or other appropriate authority whereby noncombatants whose lives are endangered by war, civil unrest, or natural disaster are evacuated from foreign countries to safe havens or to the United States.
NET	not earlier than

NGO	nongovernmental organization
NICKY	evasion and recovery contact point report (message format)
NIMA	National Imagery and Mapping Agency
NIPRNET	Nonsecure Internet Protocol Router Network
NLT	not later than
no	number
noncombatant evacuees	1. U.S. citizens who may be ordered to evacuate by competent authority include: a. Civilian employees of all agencies of the U.S. Government and their dependents, except as noted in 2a below. b. Military personnel of the U.S. Armed Forces specifically designated for evacuation as noncombatants. c. Dependents of members of the U.S. Armed Forces. 2. U.S. (and non-U.S.) citizens who may be authorized or assisted (but not necessarily ordered to evacuate) by competent authority include: a. Civilian employees of U.S. Government agencies and their dependents, who are residents in the country concerned on their own volition, but express the willingness to be evacuated. b. Private U.S. citizens and their dependents. c. Military personnel and dependents of members of the U.S. Armed Forces outlined in 1c above, short of an ordered evacuation. d. Designated aliens, including dependents of persons listed in 1a through 1c above, as prescribed by the DOS.
NORDO	no radio
NRO	National Reconnaissance Office
NSA	National Security Agency
NSOC	National Signals Intelligence (SIGINT) Operations Center
NVA	North Vietnamese Army
NVD	night vision device
OAKOC	observation and fields of fire, avenues of approach, key terrain, obstacles, and cover and concealment
OB	order of battle
OCE	operational control element
OCE(-)	operational control element (minus)
OGA	other government agency
OICC	operational intelligence coordination center
OPCEN	operations center
OPCON	**operational control**—Transferable command authority that may be exercised by commanders at any echelon at or below the level of combatant command. Operational control is inherent in Combatant Command (command authority). Operational control may be delegated and is the authority to perform those functions

of command over subordinate forces involving organizing and employing commands and forces, assigning tasks, designating objectives, and giving authoritative direction necessary to accomplish the mission. Operational control includes authoritative direction over all aspects of military operations and joint training necessary to accomplish missions assigned to the command. Operational control should be exercised through the commanders of subordinate organizations. Normally this authority is exercised through a subordinate joint force commanders and Service and/or functional component commanders. Operational control normally provides full authority to organize commands and forces and to employ those forces as the commander in operational control considers necessary to accomplish assigned missions. Operational control does not, in and of itself, include authoritative direction for logistics or matters of administration, discipline, internal organization, or unit training. (JP 3-50.2)

OPLAN operation plan—Any plan, except for the Single Integrated Operation Plan, for the conduct of military operations. Plans are prepared by combatant commanders in response to requirements established by the Chairman of the Joint Chiefs of Staff and by commanders of subordinate commands in response to requirements tasked by the establishing unified commander. Operation plans are prepared in either a complete format (OPLAN) or as a concept plan (CONPLAN). The CONPLAN can be published with or without a time-phased force and deployment data (TPFDD) file. a. OPLAN—An operation plan for the conduct of joint operations that can be used as a basis for development of an operation order (OPORD). An OPLAN identifies the forces and supplies required to execute the CINC's Strategic Concept and a movement schedule of these resources to the theater of operations. The forces and supplies are identified in TPFDD files. OPLANs will include all phases of the tasked operation. The plan is prepared with the appropriate annexes, appendixes, and TPFDD files as described in the Joint Operation Planning and Execution System manuals containing planning policies, procedures, and formats. b. CONPLAN—An operation plan in an abbreviated format that would require considerable expansion or alteration to convert it into an OPLAN or OPORD. A CONPLAN contains the CINC's Strategic Concept and those annexes and appendixes deemed necessary by the combatant commander to complete planning. Generally, detailed support requirements are not calculated and TPFDD files are not prepared. c. CONPLAN and TPFDD—A CONPLAN with TPFDD is the same as a CONPLAN except that it requires more detailed planning for phased deployment of forces. (JP 1-02)

NOTE: The term "CINC" is now used only when referring to the President of the United States. In the OPLAN definition, CINC refers to the geographic combatant commander.

OPORD operation order

OPR	office of primary responsibility
OPSEC	**operations security**—A process of identifying critical information and subsequently analyzing friendly actions attendant to military operations and other activities to: a. Identify those actions that can be observed by adversary intelligence systems. b. Determine indicators hostile intelligence systems might obtain that could be interpreted or pieced together to derive critical information in time to be useful to adversaries. c. Select and execute measures that eliminate or reduce to an acceptable level the vulnerabilities of friendly actions to adversary exploitation.
OSC	**on-scene commander**— The person designated to coordinate the rescue efforts at the rescue site. (JP 3-50.2)
OSS	Office of Strategic Services
overseas search and rescue region	Overseas unified command areas (or portions thereof not included within the inland region or the maritime region).
overt operation	An operation conducted openly, without concealment. See also clandestine operation; covert operation.
PACE	primary, alternate, contingency, and emergency. A planning acronym used in all aspects of SF mission planning and operations to ensure mission success by designating at least four plans or methods of achieving the desired end state.
pam	pamphlet
PAO	Public Affairs Office, public affairs officer
pararescue team	Specially trained personnel qualified to penetrate to the site of an incident by land or parachute, render medical aid, accomplish survival methods, and rescue survivors. (JP 3-50.2)
pax	passenger(s)
personal locator beacon	An emergency radio locator beacon with a two-way speech facility carried by crew members, either on their person or in their survival equipment, and capable of providing homing signals to assist search and rescue operations. See also emergency locator beacon; crash locator beacon. (JP 3-50.2)
PIR	priority intelligence requirements
PJ	pararescuemen
PL	phase line
PLS	**personal locator system**—lightweight airborne recovery system (LARS) equipment capable of homing to signals emitted by PRC-112 survival radios to give helicopters the range and bearing and authenticate the code in the radio. This equipment may also be known by the term "Downed Aircrew Locator System" (DALS).
POC	point of contact

pointee-talkee	A language aid containing selected phrases in English opposite a translation in a foreign language. It is used by pointing to appropriate phrases.
PR	**personnel recovery**—The aggregation of military, civil, and political efforts to recover captured, detained, evading, isolated or missing personnel from uncertain or hostile environments and denied areas. Personnel recovery may occur through military action, action by nongovernmental organizations, other U.S. Government-approved action, and diplomatic initiatives, or through any combination of these options. Although personnel recovery may occur during noncombatant evacuation operations (NEOs), NEO is not a subset of personnel recovery. (DODD 2310.2)
precautionary search and rescue/combat search and rescue	The planning and pre-positioning of aircraft, ships, or ground forces and facilities before an operation to provide search and rescue or combat search and rescue assistance if needed. The planning of precautionary search and rescue or combat search and rescue is usually done by plans personnel with search and rescue or combat search and rescue expertise and background on a J-3 (operations) staff, a joint search and rescue center, or a rescue coordination center. Also called precautionary SAR/CSAR.
PRI	primary (EPA)
PRMS	Personnel Recovery Mission Software
PRP	predesignated recovery point
PSYOP	**psychological operations**—Planned operations to convey selected information and indicators to foreign audiences to influence their emotions, motives, objective reasoning, and ultimately the behavior of foreign governments, organizations, groups, and individuals. The purpose of psychological operations is to induce or reinforce foreign attitudes and behavior favorable to the originator's objectives. (JP 1-02)
PW	prisoner of war
PZ	pickup zone
QUART	evasion and recovery handover and crossover report (message format)
RAF	Royal Air Force (United Kingdom)
raid	An operation, usually small scale, involving a swift penetration of hostile territory to secure information, confuse the enemy, or to destroy installations. It ends with a planned withdrawal upon completion of the assigned mission.
Rangers	Rapidly deployable airborne light infantry organized and trained to conduct highly complex joint DA operations in coordination with or in support of SO units of all Services. Rangers also can execute DA operations in support of conventional nonspecial operations missions conducted by a combatant commander and

can operate as conventional light infantry when properly augmented with other elements of combined arms.

RAS   **recovery activation signal**—In evasion and recovery operations, a precoordinated signal from an evader to a receiving or observing source that indicates his presence in an area and also indicates "I am here, start initial recovery planning."

RCC   **rescue coordination center**—A primary search and rescue facility suitably staffed by supervisory personnel and equipped for coordinating and controlling search and rescue and/or combat search and rescue operations. The facility is operated unilaterally by personnel of a single Service or component. For Navy component operations, this facility may be called a rescue coordination team (RCT). (JP 3-50.2)

RDZ   resupply drop zone

recovery   In evasion and recovery operations, the return of evaders to friendly control, either with or without assistance, as the result of planning, operations, and individual actions on the part of recovery planners, conventional/unconventional recovery forces, and/or the evaders themselves. (JP 3-50.2)

recovery force   In evasion and recovery operations, an organization consisting of personnel and equipment with a mission of seeking out evaders, contacting them, and returning them to friendly control.

recovery operations   Operations conducted to search for, locate, identify, rescue, and return personnel, sensitive equipment, or items critical to national security.

recovery site   In evasion and escape usage, an area from which an evader or an escapee can be evacuated.

recovery vehicle   Any land, sea, or air vehicle that performs the actual extraction of the personnel from their isolating circumstances. Also known as RV.

RESCAP   **rescue combat air patrol**—An aircraft patrol provided over a combat search and rescue objective area for the purpose of intercepting and destroying hostile aircraft. Its primary mission is to protect the search and rescue task force during recovery operations. See also combat air patrol. (JP 3-50.2)

RESCORT   **rescue escort**—Fighters or armed helicopters assigned to protect the SARTF against ground-based threats during ingress, recovery, and egress.

rescue mission commander   Serves as an extension of the JSRC coordinating SAR/CSAR efforts between the SARTF elements and the JSRC, and monitors the status of SARTF elements. Also known as RMC.

restricted operations area   Airspace of defined dimensions, designated by the airspace control authority, in response to specific operational situations/requirements within which the operation of one or more airspace users is restricted.

RFA	restrictive fire area
RFI	request for information
RGR	Rangers
RM	**recovery mechanism** (Formerly known as evasion and escape net.)—Designated infrastructure in enemy-held or hostile areas that is trained and directed to contact, authenticate, support, move, and exfiltrate U.S. military and other designated personnel to friendly control through established indigenous or surrogate networks operating in a clandestine or covert manner. RMs include, but are not limited to, unconventional assisted recovery mechanisms and may involve the use of a recovery team. RM replaces the DOD term and definition of E&E Nets. (DODI 2310.6)
ROE	rules of engagement
ROKUS	Republic of Korea-United States
RON	remain overnight
ROZ	restricted operations zone
RT	**recovery team**—An entity, group of entities, or organizations designated, trained, and directed to operate in an overt, covert, or clandestine manner in enemy-held or hostile areas for a specified period to contact, authenticate, support, move, and exfiltrate U.S. military and other designated personnel to friendly control.
RV	rendezvous
S-2	intelligence officer
S-3	battalion or brigade operations staff officer
S-4	battalion or brigade logistics staff officer
S-5	civil-military officer
SAFE	**selected area for evasion**—A designated area in hostile territory that offers evaders or escapees a reasonable chance of avoiding capture and of surviving until they can be evacuated.
SAFER	evasion and recovery selected area for evasion (SAFE) area activation request (message format)
SAID	**selected area for evasion (SAFE) area intelligence description**—In evasion and recovery operations, an in-depth, all-source evasion study designed to assist the recovery of military personnel from a selected area for evasion under hostile conditions.
SAM	surface-to-air missile
SANDY	Call sign for a USAF pilot specially trained in search procedures, aircrew survival and authentication techniques, and helicopter support tactics. (This term and its definition are applicable only

in the context of this publication and cannot be referenced outside this publication.)

SAR	**search and rescue**—The use of aircraft, surface craft, submarines, specialized rescue teams, and equipment to search for and rescue personnel in distress on land or at sea. (DOD) (JP 3-50.2)
SARDOT	search and rescue dot (on map)
SARIR	search and rescue incident report
SARNEG	search and rescue numerical encryption grid
SARREQ	search and rescue request
SARSAT	search and rescue satellite-aided tracking
SARSIT	search and rescue situation report
SARSOP	search and rescue standing operating procedures
SARTF	search and rescue task force
SATCOM	satellite communications
SAV SER SUP	Signal/Audio/Visual Service Supplement
SCAME	source, content, audience, media, and effects
SCI	sensitive compartmented information
SEAD	suppression of enemy air defenses
SEAL	sea-air-land team
search and rescue alert notice	An alerting message used for U.S. domestic flights. It corresponds to the declaration of the alert phase. Also called ALNOT.
search and rescue incident classification	Three emergency phases into which an incident may be classified or progress, according to the seriousness of the incident and its requirement for rescue service: a. Uncertainty Phase—Doubt exists as to the safety of a craft or person because of knowledge of possible difficulties or because of lack of information concerning progress or position. b. Alert Phase—Apprehension exists for the safety of a craft or person because of definite information that serious difficulties exist that do not amount to a distress or because of a continued lack of information concerning progress or position. c. Distress Phase—Immediate assistance is required by a craft or person because of being threatened by grave or imminent danger or because of continued lack of information concerning progress or position after procedures for the alert phase have been executed.
search and rescue mission coordinator	The designated person or organization selected to direct and coordinate support for a specific search and rescue mission. Also called SAR mission coordinator.

search and rescue region	See inland search and rescue region; maritime search and rescue region; overseas search and rescue region.
search mission	In air operations, an air reconnaissance by one or more aircraft dispatched to locate an object or objects known or suspected to be in a specific area.
search radius	In SAR operations, a radius centered on a datum point having a length equal to the total probable error plus an additional safety length to ensure a greater than 50 percent probability that the target is in the search area.
SecDef	Secretary of Defense
SERE	survival, evasion, resistance, escape
SERER	survival, evasion, resistance, escape, recovery
SF	Special Forces
SFC	sergeant first class
SFG(A)	Special Forces group (airborne)
SFOB	Special Forces operational base
SFODA	Special Forces operational detachment A
SIGCEN	signal center
SIPRNET	Secret Internet Protocol Router Network
SJA	Staff Judge Advocate
SMU	special mission unit
SO	special operations
SOAR	special operations aviation regiment
SOC	special operations command
SOCCE	special operations command control element
SOCCENT	Special Operations Command, United States Central Command
SOCEUR	Special Operations Command, United States European Command
SOCKOR	Special Operations Command, Korea
SOCPAC	Special Operations Command, United States Pacific Command
SOCSOUTH	Special Operations Command, United States Southern Command
SODARS	special operations debrief and retrieval system
SOF	special operations forces
SOFA	status-of-forces agreement
SOI	signal operating instructions
SOP	standing operating procedure
SPA	special psychological operations assessment

Special Forces personnel recovery coordinator	Designated individual at SFOB or FOB level who serves as commander's PR SME, responsible for the supervision and coordination of all unit PR matters. Duties include conducting PR brief at the staff mission briefing, reviewing all EPAs and ISOPREPs for completeness, assisting SFODAs with EPA preparation, maintaining continuity with the theater PR infrastructure, and serving as the PR advisor to the OPCEN staff.
SPINS	special instructions
SR	special reconnaissance
SSG	staff sergeant
SSN	social security number (EPA cover sheet)
STT	special tactics team
STU-III	secure telephone unit III
SUPCEN	support center
support forces	Support forces are any land, sea, or air forces that perform a combat mission in support of the actual extraction of the personnel from their isolating circumstances. Examples of support forces include RESCAP, RESCORT, and refueling aircraft; EW assets; ships; attack helicopters; forward air controllers; and SOF.
TACON	**tactical control**—The detailed and, usually, local direction and control of movements or maneuvers necessary to accomplish missions or tasks assigned. (JP 3-50.2)
tactical operations center	A physical grouping of those elements of an Army General and special staff concerned with the current tactical operations and the tactical support thereof. Also known as TOC.
TASKORD	tasking order
TBD	to be determined
THREATCON	terrorist threat condition
TOT	time on target
TRP	target reference point
TS	top secret
TTP	tactics, techniques, and procedures
TV	television
UAR	**unconventional assisted recovery**—NAR conducted by special operations forces (SOF). (10 USC and evolving Joint and Service doctrine for SOF define their activities with regards to NAR as UAR.) (DODI 2310.6)
UARCC	**unconventional assisted recovery coordination center**—The compartmented Special Operations Forces (SOF) facility

responsible for integration, coordination, synchronization, and deconfliction of NAR operations.

UARM **unconventional assisted recovery mechanism**—Encompasses SOF activities related to the creation, coordination, supervision, command and control, and use of recovery mechanisms either in support of combatant commands, or as directed by the NCA. UARMs may involve using an unconventional assisted recovery team. (DODI 2310.6)

NOTE: The term "NCA" is no longer used. Depending on the situation, NCA has been replaced by President, Secretary of Defense, President and Secretary of Defense, or President or Secretary of Defense. In the UARM definition, NCA indicates President or Secretary of Defense.

UART **unconventional assisted recovery team**—A designated SOF RT that is trained and equipped to operate for a specified period in hostile territory in support of personnel recovery. (DODI 2310.6)

UAV	unmanned aerial vehicle
UHF	ultra-high frequency
unconventional recovery operation	Evader recovery operations conducted by unconventional forces.
UNDER	cache report (message format)
UNPIK	United Nations Partisan Infantry, Korea
U.S.	United States
USA	United States Army
USAF	United States Air Force
USAJFKSWCS	United States Army John F. Kennedy Special Warfare Center and School
USASOC	United States Army Special Operations Command
USC	United States Code
USCENTAF	United States Central Command Air Forces
USCENTCOM	United States Central Command
USCG	United States Coast Guard
USEUCOM	United States European Command
USFK	United States Forces, Korea
USN	United States Navy
USSOCOM	United States Special Operations Command
UTM	universal transverse mercator
UW	unconventional warfare

UWOA	unconventional warfare operating area
VHF	very high frequency
WW II	World War II
WX	weather

Bibliography

The following references are cited in this publication.

AFSOC Instruction 10-3001. *Personnel Recovery*. 1 February 2002.

Bravo Two Zero, McNabb, Andy, Bantam Press, 1993.

CJCSI 3121.01. *Standing Rules of Engagement for U.S. Forces (SROE)*. 1 October 1994.

DD Form 1833. *Isolated Personnel Report*. February 1984.

Director of Central Intelligence Directive 5/1. 19 December 1984.

DODD 1300.7. *Training and Education to Support the Code of Conduct (CoC)*. 8 December 2000.

DODD 2310.2. *Personnel Recovery*. 22 December 2000.

DODI 1300.21. *Code of Conduct (CoC) Training and Education*. 8 January 2001.

DODI 2310.4. *Registration of Prisoners of War (POW, Hostages, Peacetime Government Detainees, and Other Missing or Isolated Personnel)*. 21 November 2000.

DODI 2310.6. *Non-Conventional Assisted Recovery in the Department of Defense*. 13 October 2000.

FM 27-10. *The Law of Land Warfare*. 18 July 1956.

JP 1-02. *Department of Defense Dictionary of Military and Associated Terms*. 12 April 2001.

JP 2-0. *Joint Doctrine for Intelligence Support to Operations*. March 2000.

JP 2-01. *Appendix C to Joint Intelligence Support to Military Operations*. 20 November 1996.

JP 2-02. *National Intelligence Support to Military Operations*. 28 September 1998.

JP 3-50.2. *Doctrine for Joint Combat Search and Rescue*. 26 January 1996.

JP 3-50.3. *Joint Doctrine for Evasion and Recovery*. 6 September 1996.

JP 3-50.21. *Joint TTP for Combat Search and Rescue*. 23 March 1998.

The One That Got Away, Ryan, Chris, Century, 1995.

USC, Title 10, Section 1501, *System for Accounting for Missing Persons*. 5 January 1999.

USC, Title 50, Section 435, *Procedures*. 1 February 2001.

USSOCOM Directive 525.21. *Personnel Recovery*. 4 May 2000.

The following references represent those that would be found in a PR planner's library. A list of web sites applicable to PR follows the references.

ACC SAR OPLAN 529. 30 January 1992 (Unclassified/ROKUS).

AF Pam 64-5. *Aircrew Survival.* 1 September 1995.

AFDD 2-1.6. *Combat Search and Rescue.* 15 July 1998.

AFDD 34. *Combat Search and Rescue Operations.* 30 December 1994.

AFDD 35. *Special Operations.* 16 January 1995.

AFI 13-208. *Rescue Coordination Center Combat Search and Rescue Operating Procedures.* 1 January 1996.

AFR 64-3. *Combat Search and Rescue Procedures.* 25 February 1985.

AFTTP 3-2.26. *Multi-service Procedure for Survival, Evasion, and Recovery.* June 1999.

Allied Tactical Publication 62 (Draft). *NATO Combat Search and Rescue.* 3 August 1998.

AR 500-2. *Search and Rescue (SAR) Operations.* 15 January 1980.

AR 525-90. *Wartime Search and Rescue (SAR) Procedures (AFR 6403; NWP 19-2). Reprinted With Basic Incl C1.* 17 March 1989.

CJCSI 3150.25A. *Joint Lessons Learned Program.* 1 August 2000.

CJCSI 3270.01. *Personnel Recovery Within the Department of Defense.* 1 July 1998. (SECRET)

CJCSM 3122.01. *Joint Operation Planning and Execution System, Volume I, Planning Policies and Procedures.*

CJCSM 3122.02. *Joint Operation Planning and Execution System, Execution Planning.*

CJCSM 3122.03. *Joint Operation Planning and Execution System, Volume II, Planning Formats and Guidance.*

CJCSM 3122.04. *Joint Operation Planning and Execution System, Volume II.*

CJCSM 3141.01A. *Procedures for the Review of Operation Plans.* 15 September 1998.

DODD 1300.7. *Training and Education to Support the Code of Conduct (CoC).* 8 December 2000.

DODD 2310.2. *Personnel Recovery.* 22 December 2000.

DODD 3025.15. *Military Assistance to Civil Authorities.* 18 February 1997.

DODD 5111.10. *Assistant Secretary of Defense for Special Operations and Low-Intensity Conflict (ASD [SO/LIC]).* 22 March 1995.

DODI 1300.18, *Military Personnel Casualty Matters, Policies, and Procedures.* 27 November 1991.

DODI 2310.3. *Personnel Recovery Response Cell Procedures (PRRC).* 6 June 1997.

DODI 2310.5. *Accounting for Missing Persons.* 31 January 2000.

DODI 2310.6. *Non-Conventional Assisted Recovery in the Department of Defense.* 13 October 2000.

DOD Memorandum of Understanding (MOU) between DOD and DOS. *Protection and Evacuation of U.S. Citizens and Designated Aliens Abroad.* 28 and 29 September 1994.

(U) DOD National Intelligence Community Support for Personnel Recovery (S). March 1999.

FM 1-100. *Army Aviation Operations.* 21 February 1997.

FM 1-111. *Aviation Brigades.* 27 October 1997.

FM 3-05.20. *Special Forces Operations.* 26 June 2001.

FM 3-05.60. *ARSOF Aviation Operations.* 15 August 2000.

FM 3-05.70. *Survival.* 17 May 2002.

FM 3-05.71. *(C) Resistance and Escape (U).* 7 September 2001.

FM 3-05.220. *(S) Special Forces Advanced Operations Techniques (U).* 30 September 1993.

FM 8-10-6. *Medical Evacuation in a Theater of Operations, Tactics, Techniques, and Procedures.* 14 April 2000.

FM 10-52. *Water Supply in Theater of Operations.* 11 July 1990.

FM 27-10. *The Law of Land Warfare.* 18 July 1956.

FM 90-31. *AMCI Army and Marine Corps Integration in Joint Operations (MCRP 3-3.8).* 29 May 1996.

FM 100-103. *Army Airspace Command and Control in a Combat Zone.* 7 October 1987.

International Aeronautical and Maritime Search and Rescue Manual (IAMSAR). 11 June 2001.

JP 0-2. *Unified Action Armed Forces (UNAAF).* July 2001.

JP 1-0. *Doctrine for Personnel Support to Joint Operations.* 19 November 1998.

JP 1-02. *Department of Defense Dictionary of Military and Associated Terms.* 12 April 2001.

JP 2-0. *Joint Doctrine for Intelligence Support to Operations.* March 2000.

JP 2-01. *Appendix C to Joint Intelligence Support to Military Operations.* 20 November 1996.

JP 2-02. *National Intelligence Support to Military Operations.* 28 September 1998.

JP 3-0. *Doctrine for Joint Operations.* 10 September 2001.

JP 3-05. *Doctrine for Joint Special Operations.* 17 April 1998.

JP 3-05.3. *Joint Special Operations Operational Procedures.* 25 August 1993.

JP 3-07. *Joint Doctrine for Military Operations Other Than War.* 16 June 1995.

JP 3-07.5. *Joint Tactics, Techniques, and Procedures for Noncombatant Evacuation Operations.* 30 September 1997.

JP 3-13.1. *Joint Doctrine for Command and Control Warfare (C2W).* 7 February 1996.

JP 3-50.2. *Doctrine for Joint Combat Search and Rescue.* 26 January 1996.

JP 3-50.3. *Joint Doctrine for Evasion and Recovery.* 6 September 1996.

JP 3-50.21. *Joint TTP for Combat Search and Rescue.* 23 March 1998.

JP 3-52. *Doctrine for Joint Airspace Control in the Combat Zone.* 22 July 1995.

JP 3-53. *Doctrine for Joint Psychological Operations.* 10 July 1996.

JP 3-55.1. *Joint Tactics, Techniques, and Procedures for Unmanned Aerial Vehicles.* 27 August 1993.

JP 3-56.1. *Command and Control for Joint Air Operations.* 14 November 1994.

JP 5-0. *Doctrine for Planning Joint Operations.* 13 April 1995.

JP 6-0. *Doctrine for Command, Control, Communications, and Computer (C4) Systems Support to Joint Operations.* 30 May 1995.

MCM 213-98. *(S) MOA Concerning DOD-CIA Mutual Support in Policy, Research and Development, Training, Planning, and Operations for Personnel Recovery (U).* 7 October 1998.

(S) NSA Personnel Recovery Concept of Operations for National Intelligence Support (U). 21 May 1998.

ROKUS OPLAN 5027-98 (S).

USCENTAF Instruction 10-101. *Personnel Recovery Procedures.* 31 January 1997. (In Revision)

USCENTCOM OPLAN 1003-98 (S).

USCENTCOM Regulation 525-10. *(S) Procedures for Processing Recovered Personnel (U).* 7 June 1995. (In Revision)

USCENTCOM Regulation 525.32. *Personnel Recovery Procedures.* 3 April 2000.

USEUCOM Directive 55-13. *Personnel Recovery.* 6 July 1998.

USFK Regulation 525-40. *(S/ROKUS) Personnel Recovery (U).* 20 May 1997.

USSOCOM Directive 525-21. *Personnel Recovery.* 4 May 2000.

NOTE: The following are SIPRNET (classified) web addresses related to PR:

Air Force Operations Group Home Page:

http://ga14.af.pentagon.smil.mil/afog

Department of Prisoner of War/Missing in Action Office (DPMO) Home Page:

http://peacock.policy.osd.pentagon.smil.mil/dpmo/

DIA Military Geography Division:

http://delphi-s.dia.smil.mil/intel/oicc/twj/twj2/links/databases-links.html

Joint Personnel Recovery Agency Home Page:

http://www.jpra.jfcom.smil.mil

JTF-SWA JSRC Home Page:

http://www.centaf.psdb.af.smil.mil

National Imagery and Mapping Agency (NIMA) Home Page:

http://www.nima.smil.mil

Naval Strike and Air Warfare Center (NSAWC) home Page:

http://www.nsawc.navy.smil.mil

Pacific Rescue Coordination Center Home Page:

http://www.cidss.hickam.af.smil.mil

United States Central Command Personnel Recovery Home Page:

http://www.centcom.smil.mil

United States European Command Home Page:

http://www.eucom.smil.mil

United States Joint Forces Command Home Page:

http://157.224.120.250

United States Special Operations Command Home Page:

http://www.socom.smil.mil

NOTE: *The following are NIPRNET (unclassified) addresses related to PR:*

AFSOC publications:

https://www.afsoc.af.mil/milonly/library1/

Joint Personnel Recovery Agency:

http://www.jpra.jfcom.mil

Many DODDs and DODIs can be found in full text at:

http://www.dtic.mil/whs/directives

Many JPs can be found in full text at:

http://www.dtic.mil/doctrine

Mountain Rescue Association:

http://www.mra.org

> Includes international, as well as U.S. SAR, links.

National Association for Search and Rescue (NASAR):

http://www.nasar.org

> Links to an on-line library and bookstore catalog, photographs, conferences, and other SAR-related links.

Index

A
architecture, 2-2 – 2-4, D-2
    ARSOTF, 2-4
    SOF, 2-3
    theater, 2-2
    UARCC, D-2

C
Civil Affairs support to PR, 5-1 – 5-2
command relationships, 2-1– 2-11
    coordination for recovery of personnel, 2-5
    launch and execute authority, 2-4
    tasking authority (JSRC), 2-5 – 2-6

D
DD Form 1833 (ISOPREP), 2-8 – 2-9
debriefing of recovered personnel, 2-23 – 2-24, 3-25

E
EPA, Appendixes A, B, C
    EPA format, Appendix A
    FOB EPA guidance format, Appendix B
    planning considerations, Appendix C
evader actions, 3-12 – 3-15
    contact procedures, 3-14
    individual responsibilities, 3-12, 3-15
    signaling, 3-13
evasion aids, 3-8 – 3-12
    blood chit, 3-11
    EVC, 3-10
    pointee-talkee, 3-11 – 3-12

survival equipment, 3-8
evasion planning information sources, 3-6 – 3-7

F
five tasks of NAR, 1-17
five tasks of PR, 1-6 – 1-8

H
history (evolution) of PR, 1-2 – 1-5

I
intel analyst, 2-15 – 2-16
ISOPREP (See DD Form 1833.)

J
JSRC, 2-5 – 2-6

M
methods, 1-12 – 1-17
    component CSAR, 1-14
    JCSAR, 1-14
    multinational, 1-16
    NAR (UAR), 1-16
    opportune, 1-14
    unassisted, 1-13

O
options, 1-12 – 1-13
    civil option, 1-12
    diplomatic option, 1-12
    military option, 1-13

P
planner checklists, E-1 – E-17
policy, 1-5 – 1-6

PSYOP support to PR, 5-2 – 5-5

R
RCC, 2-7
recovery considerations, 1-8 – 1-11

S
SARIR, 2-21
SARSIT, 2-21
SF PR cell, 2-10
SF PR coordinator, 2-11
SF PR missions, 3-16 – 3-45
    component (unilateral) recovery, 3-18 – 3-19
    opportune support to PR, 3-16 – 3-18
    support to joint recovery operations, 3-19 – 3-23
    support to multinational PR operations, 3-24 – 3-25
    UAR, 3-25 – 3-45

T
training, 3-3 – 3-5

U
UAR
    assets, 3-43 – 3-45
    definition, 3-25
    planning, 3-42 – 3-43
    specified tasks, 3-26 – 3-34
    RT, 3-34
UARCC
    command authority (JFSOCC), D-2
    manning, D-3
    mission, D-1

organization, Figure D-1
staff roles and
 responsibilities, D-3
tasks, 2-10, D-1
UARM, 3-37 – 3-38,
 Figure 3-21
UART, 3-34, Figure 3-18
unassisted evasion, 3-1 – 3-15

FM 3-05-231
13 JUNE 2003

By Order of the Secretary of the Army:

ERIC K. SHINSEKI
General, United States Army
Chief of Staff

Official:

JOEL B. HUDSON
Administrative Assistant to the
Secretary of the Army
0319608

DISTRIBUTION:

Active Army, Army National Guard, and US Army Reserve: To be distributed in accordance with the initial distribution number 115902, requirements for FM 3-05.231.

PIN: 080927-000

www.ingramcontent.com/pod-product-compliance
Lightning Source LLC
Chambersburg PA
CBHW050057230526
45470CB00004B/1562